Advances in Planetary Science – Vol. 1

NUCLEAR PLANETARY SCIENCE

Planetary Science Based on Gamma-Ray, Neutron and X-Ray Spectroscopy

Advances in Planetary Science

Series Editor: Wing-Huen Ip *(National Central University, Taiwan)*

Published

Advances in Planetary Science – Vol. 1

NUCLEAR PLANETARY SCIENCE

Planetary Science Based on Gamma-Ray, Neutron and X-Ray Spectroscopy

Nobuyuki Hasebe
Waseda University, Japan

Kyeong Ja Kim
University of Science and Technology, Korea
Korea Institute of Geoscience and Mineral Resources, Korea

Eido Shibamura
Waseda University, Japan

Kunitomo Sakurai
Kanagawa University, Japan & Waseda University, Japan

World Scientific

NEW JERSEY · LONDON · SINGAPORE · BEIJING · SHANGHAI · HONG KONG · TAIPEI · CHENNAI · TOKYO

Published by

World Scientific Publishing Co. Pte. Ltd.

5 Toh Tuck Link, Singapore 596224

USA office: 27 Warren Street, Suite 401-402, Hackensack, NJ 07601

UK office: 57 Shelton Street, Covent Garden, London WC2H 9HE

Library of Congress Control Number: 2017950256

British Library Cataloguing-in-Publication Data
A catalogue record for this book is available from the British Library.

Advances in Planetary Science — Vol. 1
NUCLEAR PLANETARY SCIENCE
Planetary Science Based on Gamma-Ray, Neutron and X-Ray Spectroscopy

ISBN 978-981-3209-70-1

Desk Editor: Amanda Yun

Typeset by Stallion Press
Email: enquiries@stallionpress.com

Preface

Nuclear geological and cosmochemical methods play various important roles in our understanding of the origin and evolution of planets and other celestial bodies in the solar system. The progress in the research on the planets is well-recognized, and is closely associated with our understanding of Earth; its age and evolution, and its relation to the origin of life and its evolution.

We, human beings as the residents of Earth, have been making extensive efforts to understand the origin and evolution of Earth, and then extending the use of the same research methods and techniques to investigate other planets and their satellites and various types of small bodies found in space. The investigation of the internal structures and environmental situation associated with the origin and evolution of various planetary bodies and their future prospects is rapidly progressing through the application of techniques to directly observe the physical and chemical properties of the planets and their satellites, with the help of artificial space probes such as the Apollo, the various Pioneers and Voyagers, and Cassini, and Curiosity.

Furthermore, the dating techniques and methods have also been successful in estimating the ages of Earth, Moon, and other planets and their satellites; and have further extended into the study of various kinds of meteorites and their debris.

Nuclear planetology, a new branch of planetary science, is especially useful for research on the present states of the structures of various planetary bodies and their satellites. This branch is making progress together with the progress of sensor and electronics technologies. Extensive reviews of these developments have been provided for readers in this book.

This book is organized as follows. Chapter 1 gives readers a general summary of what current science knows about the origins of the universe and various types of celestial bodies; and introduces readers to planetary science, and how the specific field of nuclear planetary science came to be. In Chapter 2, we describe the principle of nuclear spectroscopy and its use in determining the elemental composition in planetary bodies. In Chapter 3, the radiation detectors that are used with nuclear spectrometers to study planetary gamma-rays, neutrons, and X-rays are explained. In Chapter 4, we present numerical simulation and the proton beam experiment for nuclear planetary explorations conducted so far. In Chapter 5, nuclear planetary spectrometers are described. Chapter 6 includes a description of the reduction and analysis of observational data obtained by nuclear spectrometers. Chapter 7 presents recent advancements in lunar, Martian, and other planetary sciences relating to nuclear planetary science. Future missions using nuclear spectroscopy for the explorations of the Moon, Martian satellites, and Near Earth Asteroids are also proposed. In Chapter 8, nuclear methods to study signs of life in the universe are described. A discussion of the implications of the results in lunar geochemistry are also included. The prospect of nuclear planetary science and the role of future missions are summarized in the final chapter.

This book allows the reader to acquire a clear understanding of the scientific fundamentals behind nuclear spectroscopic technologies that will be used on planets and asteroids in the future. This book may also be useful for industry developers interested in joining national and international space programs. Finally, it may be used for undergraduate, postgraduate students and young researchers in natural sciences and engineering.

Since the initial submission of this manuscript to the publisher, the work has been carefully treated and improved on by the staff of World Scientific. We would like to specially thank them — especially Senior Editor, Ms. Amanda Yun — for making this manuscript much more readable and understandable.

We also express our sincere thanks to Dr. Ip-Wing-Huen at the Graduate Institute of Astronomy, National Central University, Taiwan, for his kind advice and suggestions for this book.

Finally, it seems reasonable and appropriate to express our appreciation to various works contributed by colleagues and researchers in the research fields considered in this book — such as the field of nuclear cosmochemical techniques, which has been established under the umbrella of Nuclear Planetary Science as a branch of astrochemistry and planetology.

We believe that the contents and interpretation expressed in this book would be well received by anyone who is considering to take up the study of the subjects covered in this book, and hope that readers will find this book highly beneficial and useful.

Nobuyuki Hasebe, Kyeong Ja Kim,
Eido Shibamura & Kunitomo Sakurai

December 2016

About the Authors

Nobuyuki HASEBE is a full Professor in the Faculty of Science and Engineering, Department of Physics/School of Advanced Science and Engineering, Department of Pure and Applied Physics/Graduate School of Advanced Science and Engineering, Research Institute for Science and Engineering, Waseda University, Japan, and director of the university's Cosmic Radiation Physics Laboratory. His research began at the Institute of Cosmic Ray Research, University of Tokyo, Japan, and then at Ehime University, Japan, for 19 years, before he joined Waseda University in 1998. He is also Principal Investigator of KAGUYA Gamma-Ray Spectrometer (GRS) on the lunar orbiter SELENE (KAGUYA). An expert in space science, radiation physics and nuclear detectors, with more than 200 papers published in science journals and proceedings, Prof Hasebe has participated in many space missions (OHZORA, GIOTTO, GEOTAIL, NOZOMI, KAGUYA, some for Shuttle and ISS experiments, etc.) and provided scientific instruments on-board spacecraft. He has supervised 14 PhDs, more than 50 MSs and more than 100 BS research projects at Waseda University. He has also organized a number of international workshops and conferences in radiation detection and lunar science, and has served as referee in many international journals. He also acts as a regular referee/adviser in several Japanese committees. His research fields are: Radiation Physics, Radiation Pertinent Physics, Cosmic Ray Physics, Space Science, and Nuclear Planetary Science.

Kyeong Ja KIM is a professor at the Department of Geophysical Exploration, University of Science and Technology, Korea, and Principal Investigator at the Korea Institute of Geoscience and Mineral Resources. A visiting scientist at the Planetary Science Branch, NASA Ames Research Centre in 2013, she was involved in data analysis of the Mars Odyssey GRS Program at the Lunar and Planetary Laboratory of University of Arizona, USA, and in the KAGUYA GRS Program as a Co-PI. She has served as co-convener and science committee member for AOGS and COSPAR for lunar science and lunar exploration sessions, and is a foreign collaborator for NASA SSERVI's FINESSE Project. Prof Kim holds 8 patents and 1 intellectual property right, and has 58 SCI papers, 28 other publications, and over 100 proceedings and abstracts under her belt. Her research fields are: Planetary payload development in nuclear planetology, cosmogenic nuclide production on planetary surface, paleoclimate change, and geochronology.

Eido SHIBAMURA was awarded his PhD by Waseda University, Japan. He engaged in liquid rare gas filled ionization detectors and obtained the ratio of diffusion coefficient to mobility for electrons, eD/μ, in liquid argon. He also studied the basic parameter of liquid rare gas filled detectors for ionizing radiation, including the energy resolution, W-value and drift velocity of electrons in the liquids. He received his PhD with this work. He studied the photon yield for gamma rays in NaI(Tl) scintillator and found that the photon yield is at least 50% larger than that in literature. Then he systematically measured the photon yield in some important inorganic scintillators on the basis of the measured effective quantum efficiency of photomultipliers coupled to the scintillators. He found that the effective quantum efficiency of the Photomultipliers for 45 degree incident photons are 20–50% larger than those given by the PMT manufacturer for normally incident photons. He was a co-investigator of the KAGUYA gamma ray spectrometer. His research interests include the detector for ionizing radiation, scintillation detectors, and those with liquefied rare gases, and the measurement of gamma-ray emitted from the surface of the Moon.

Kunitomo SAKURAI is Professor Emeritus, Kanagawa University, Japan, and a Research Scientist at the Research Institute for Science and Engineering, Waseda University, Japan. His research began at the Geophysical Institute, followed by the Ionosphere Research Laboratory at Kyoto University, Japan. He then took a job at the Goddard Space Flight Center, NASA, USA, for almost eight years, before taking on a professorship at Kanagawa University, Japan. While working there, Prof Emeritus Sakurai was a member of the Faculty of Engineering for six years, and the dean of the faculty as well as the President of the University for three years. After retiring from Kanagawa University, he joined the Research Institute for Science and Engineering, Waseda University, in his current position since 2004. His main areas of research over the years are summarized as follows: Galactic and Solar Cosmic Rays and Solar Neutrino flux from the Sun, and the "Sakurai Periodicity" (the technical term given on the quasi biennial variation seems in the flux).

Contents

Chapter 1

Introduction

The creation of the solar system is a part of the history of the universe. And the universe's story began at its creation and continues on into the future. Space and planetary sciences have clearly shown how the formation and evolution processes materialized — through the use of various ideas and advanced measurement technologies, observational astronomy, and/or numerical simulations. These researches are related to how the present global environment came to be, as well as the most needed recourse of our time: water — the existence of water and how it sustains living beings on Earth. All in all, these studies help us find answers to the most important question of our time: how we can protect and continue to maintain life on Earth.

The question of the origin and evolution of our solar system should be approached from various fields, such as astronomy, physics, chemistry, petrology, mineralogy, geology, meteorology, and biology. Efforts by researchers in each field to try to understand and clarify the truth have increased. Using approaches based on physics or chemistry, we try to understand the origin of the solar system and the Earth by studying the nature of atoms or nuclei that constitute the materials in the solar system, and by combining this knowledge with that from other fields.

1.1 Planetary Science

Our solar system consists of the Sun at its gravitational center and a planetary system of eight planets, their natural satellites, many dwarf planets, as well as a large number of tiny objects — all are bound by gravity and orbit the Sun. It formed about 4.54 billion years ago from the gravitational collapse of a giant molecular cloud. In comparison to the universe, which formed around 13.8 billion years ago, our solar system is very young. It is located close to the inner rim in the Local Bubble, about halfway along the Orion Arm, or 8 kpc from the Galactic Center. The nearest star to our solar system is α Centauri, which is a group of three stars, α Centauri A, α Centauri B, and Proxima, approximately 1.3 pc from the Sun.

Stars, including our Sun, are formed from relatively dense molecular clouds in the galaxy that have evolved through rather violent stages (e.g. T Tauri) before settling down on the main sequence. Star formation, at least for stars of the solar mass, is not a quiet gravitational collapse from a gas cloud; instead, their formation is characterized by highly energetic mass outflows. Protostars may lose large amounts of mass in very short timescales ($\sim 10^5$ years). At this stage, the central star (surface temperature ~ 4000 K) becomes visible, surrounded by a nebular disk. In the next stage, strong bipolar outflows of gas and jets of material break out, mainly along the rotation axis of the star, which is a characteristic of the evolution of young stars. Infalling gas does not fall directly on the rotating protostar object but forms a disk moving around the protostar, perpendicular to the rotation axis.

More and more infalling materials accrete to the disk rather than to the protostar. Radiation from the disk adds excess infrared energy to the energy spectrum showing a broad emission hump with absorption at 10 μm from silicate grains and 3.1 μm from water ice. Finally, the disk disappears and a planetary system may form from the inner planets to the outer planets.

The four small inner planets (also called the terrestrial planets) of our solar system, Mercury, Venus, Earth, and Mars, are primarily composed of rock and metal. Among the four outer planets, the largest are Jupiter and Saturn. They are called the "gas planets" or "gas giants". Composed mainly of hydrogen and helium, they are substantially more massive than the inner or terrestrial planets. The two outermost planets are Uranus and Neptune. They are called "ice planets" or "ice giants" and are mostly made up of "ices" such as frozen water, ammonia, and methane, because they are farthest from the Sun.

The solar system also contains a large number of small objects. The asteroid belt (main belt) that lies between the orbits of Mars and Jupiter, is similar to the terrestrial planets as it mostly contains objects composed of rock and metal. Populations of trans-Neptunian objects lying beyond Neptune's orbit are called the Kuiper belt and these objects are composed mostly of ices. Within these populations, there are dwarf planets that may be large enough to have been rounded by their own gravity. At present, dwarf planets that have been identified include the asteroid Ceres and the trans-Neptunian objects Pluto, Eris, Haumea, and Makemake. In addition, various other small-body populations including rings and interplanetary dust orbit in the solar system.

In the next section, we describe the terrestrial planets, asteroids, and comets.

1.1.1 *Planetary bodies*

1.1.1.1 *Terrestrial planets*

The terrestrial planets of our solar system consist of four planets, as mentioned earlier. Among them, only one terrestrial planet, Earth, is known to have an active hydrosphere. During the formation of the solar system, there were probably many more "terrestrial" planetesimals, but they have all either merged with protoplanets, or have been ejected away, and these four planets have remained in the solar nebula.

A terrestrial planet, or rocky planet, is composed primarily of silicate rocks and metals. Hydrogen compounds, such as water and methane, typically condense at low temperatures and remain gaseous inside the frost line where temperatures are high. The rocky and metallic materials are better suited to condense at high temperatures. Thus, the inner planets are made mostly of rock and metal, and are thus defined as sterrestrial planets. The terrestrial planets are the inner planets closest to the Sun. They have a solid planetary surface, making them substantially different from the much larger gas giants, which are composed mostly of some elemental combination of H, He, and volatiles. All terrestrial planets have approximately the same structure: A central metallic core, consisting mostly of elements, Fe and Ni, with a surrounding silicate mantle. Our Moon is quite similar, though it has a much smaller iron core. Terrestrial planets can have a large number of craters, mountains, volcanoes, and other surface structures. They have secondary atmospheres, generated through the activity of internal volcanism or comet impacts, in contrast to the gas giants, which only have primary atmospheres captured directly from the original solar nebula.

1.1.1.2 *Gas planets and icy planets*

Outside the frost line, temperatures are cooler and hydrogen compounds are able to condense into ices. Rock and metal are still present in the outer solar system but both are outnumbered and outweighed by the hydrogen compounds. Thus, the planetesimals formed in the outer solar system are composed primarily of hydrogen compounds with traces of rock and metal. Gas planets are referred to as gas giants, giant planets, or Jovian planets. The two gas planets in our solar system are Jupiter and Saturn. They are mostly composed of H and He. The outermost planets are called "ice giants". Uranus and Neptune are the two ice giants in our solar system. Although ice giants do also have some H and He, they are mostly made up of "ices" such as CH_4, NH_3, and H_2O. Methane gives Uranus and Neptune their blue color.

1.1.1.3 *Asteroids, comets, and others*

Asteroids essentially consist of many small bodies made of rocks in the size of 1 m–10 km and they are thought to be the building blocks of planets and/or their satellites. The largest asteroid, Ceres, is about 950 km wide. Most of the asteroids are located in the asteroid belt (called the "main belt") between the orbits of Mars and Jupiter. Some of the main-belt materials are primordial material that has never differentiated before. Other asteroids are a parts of planetary bodies that were broken apart by a collision during the formation stage of planets. These asteroids are thought to be linked with meteorites. Therefore, the exploration of these tiny asteroids is closely associated with the study on how the solar system formed and evolved.

Some of numerous asteroids that approach or cross the Earth's orbit are called near-Earth objects (NEOs). They are not only an important target of scientific study but also the target of space exploration and development because they represent a huge storage of natural resources free of Earth's gravity. NEOs can supply materials for a wide range of operations both in space and on Earth, as they are thought to contain large amounts of water, carbon, structural metals, industrial material, and precious metals.

A "comet" is an icy small body made of ice and rock. As comets with highly elliptical orbits approach the Sun, they are heated up and then begin to outgas, and they sometimes form "comas" and long tails. Comets are distinguished from asteroids by the presence of a gravitationally unbounded and extended atmosphere surrounding their central nucleus, or coma, when they approach the Sun. Short-period comets such as Halley's Comet were perturbed out from the Kuiper belt beyond Neptune's orbit. Long-period comets may come from the Oort Cloud.

1.1.2 *Formation and evolution of planets*

1.1.2.1 *Planetary formation*

The solar system was formed from a large gas nebula that had dust grains in it. The nebula collapsed under its own gravity to form the Sun and the (known) eight planets. The planets in the solar system are thought to have formed from the disc-shaped cloud of gas and dust left over from Sun's formation. The planets began as dust grains in orbit moving around the protostar. These grains collided to form larger bodies of 100 m in size, and further collisions gradually increased their sizes to those of planetesimals of ~1 km. The families of small bodies consist of comets and asteroids, ring particles, and satellites.

The inner solar system, inside 4 AU, was too warm for gases and volatile molecules like water and methane to condense, such that the planetesimals formed there could only form from compounds with high melting points such as metals (like Fe, Ni, Mg, Al, and Ca), and rocky silicates. These rocky bodies would become the terrestrial planets. The initial terrestrial planets grew to about 0.05–0.1 Earth masses (Mars-size bodies) and ceased accumulating matter about 10^5 years after the formation of the Sun. Subsequent collisions between these planet-sized bodies allowed terrestrial planets to grow to their present sizes. The asteroid belt between the orbits of Mars and Jupiter is made of billions of objects of all sizes from microscopic grains up to large bodies measuring a few hundreds of kilometers. Through the collisions and impacts, debris are ejected into the solar system. The materials colliding with Earth consist of the dominant source of meteorites, which in turn are unique extraterrestrial materials sampling their asteroid parent bodies. In fact, the few thousand samples of meteorites so far collected on Earth show a wide variety of properties, allowing them to be classified in a few families, matching the diversity of the asteroids from which they originate. Some meteorites are known as carbonaceous chondrites that contain high abundances of volatile species, reflecting their origin from the least evolved objects. The study opens up the possibility to probe the pristine material from which the inner planets were formed.

1.1.2.2 *Evolution of planets*

The release of gravitational energy during the accretion process results in hot molten planetary bodies called the magma ocean. After a terrestrial planet has formed, the initial differentiation into a rocky mantle and a metallic core causes element fractionation based on the relative effects of their siderophile, chalcophile, and lithophile affinities. The composition of the silicate mantle is dominated by the geochemistry of lithophile elements that form ionic bonds with oxygen.

The lithophile elements are often divided into "compatible" and "incompatible" elements. In geochemistry, compatibility is a measure of how readily a particular trace element substitutes for a major element within a mineral. The compatibility of an ion is controlled by its valence and its ionic radius. An element is incompatible when it is unsuitable in size and/or charge to the cation sites of the minerals. During the fractional crystallization of magma, elements that have difficulty in entering cation sites of the minerals are concentrated in the melting phase of the magma. Compatible elements are those that enter the Fe and Mg sites in mantle minerals. Elements that are excluded from the dominant mantle minerals are termed "incompatible". During the formation and crystallization of magmas, such elements, including radiogenic elements such as K, Th, and U, are concentrated and so eventually find their way into the crust. KREEP

(K, potassium; REE, rare-earth elements; and P, phosphorus) is a geochemical component of lunar impact breccias and basaltic rocks. Its most distinguished feature is the enhanced concentration of incompatible elements and the heat-generating elements, namely radioactive U, Th, and K.

The cooling process of the planets drives the tectonic processes that form a variety of planetary surface features. One of the characteristics is the formation of planetary crusts that are distinct from the bulk composition of the planets. The formation of crusts/mantles leads to the concentration of incompatible elements, and the processes accelerate mantle cooling because the incompatible elements include the radiogenic heat-producing elements K, Th, and U. The crusts on rocky planets are divided into three types: primary crusts, secondary crusts, and tertiary crusts.

Primary crusts are formed during initial planetary differentiation from the melting phase, which is formed in a short timescale, followed by accretion. One of the distinguishing features of these crusts is that they contain low concentrations of incompatible elements because primary crusts are formed from mostly molten planets. The heavily cratered ancient crust of Mars in the southern highlands and the highland crust of the Moon were probably created directly following the formation of the planetary bodies.

Secondary crusts form on longer timescales. These products of the partial melting of the silicate mantles of rocky planets include various species of basalts. Typical examples are the earth's oceanic crust, the surface of Venus, the volcanic surface of the Tharsis plateau, the northern plains on Mars, and the maria of the Moon. The basalts forming the dark maria on the Moon were derived from the partial melting of a limited volume of mantles from its interior.

Tertiary crusts are formed through the dehydration or melting of secondary crusts. Earth's continental crust remains the only current example in our solar system. They can contain enrichments of incompatible elements resembling primary crusts due to the continuous recycling of the secondary crust, because such crusts build up over long periods of time. The formation of crusts results in enrichments of incompatible elements.

Earth's interior is composed of several layers of varying density and composition. The structure of the earth can be divided into five parts: the lithosphere, asthenosphere, mesospheric mantle, outer core, and inner core. And the interior of the earth is chemically divided into five important layers: the crust, upper mantle, lower mantle, outer core, and inner core. The topmost is the crust with a 5–70 km thick layer of Si, Al, and Mg. Earth's mantle extends to a depth of 2,890 km. The upper mantle is divided into the lithospheric mantle and the asthenosphere. The upper mantle and lower mantle are separated by the transition zone. The lowest part of the mantle next to the core–mantle boundary is known as the D'' layer. Seismic measurements tell us that the core is divided into two parts, a liquid outer core extending to a radius of ~3,400 km and a solid inner core with a radius of ~1,220 km. The densities are between 9.9 and 12.2 g/cm^3 in the outer core and 12.6 and 13.0 g/cm^3 in the inner core. The liquid outer core surrounds the inner core and is believed to be composed of Fe mixed with Ni and traceable amounts of lighter elements. Recent speculation suggests that the innermost part of the core is composed of Au, Pt, and other siderophile elements.

Investigation of the internal structure of Earth includes the chemical composition of the cores (inner and outer) and the convective motion (as a superplume inside the mantle), whereas the investigation of the physical properties of Earth's outer core includes the study of the presence of light elements such as oxygen, silicon, and other refractory lithophile elements. These elements may have been uncovered by the superplume motion deep inside the mantle and then transferred to the outer core. This motion seems to play an important role in the transport of these elements to the outer core. So, according to current understanding, oxygen, silicon, and other refractory lithophile elements are continuously accumulated in the outer core.

Earth is composed of three different tectonic domains: the subsurface domain (where plate tectonics happens), the intermediate lower mantle domain (where plume tectonics takes place), and the core (where growth tectonics occurs) (Yuen *et al.*, 2007). Tectonic plates supply cold materials to the mantle transition zone at 670 km in depth. The cold plume formed flows down to the outer core to refrigerate the metallic Fe–Ni liquid there, to initiate a new downflow in the core. The super-upwelling of mantle flow seems to be a passive response of cold plumes, which is important in the Earth's dynamics (Isozaki, 2009; Maruyama, 1994). The uprising super-plume caused the continental break-up and subsequent dispersion.

Plate tectonics is a superficial phenomenon on Earth in the region that is less than one-tenth of Earth's radius; where large-scale plume flows dominate the major parts of deeper mantle. Thermal-material mantle convections, the breaking-up of super cratons in the tectosphere, and the convection patterns of the outer core, are all primarily caused by cold plumes rather than hot mantle upwelling (Isozaki, 2009; Maruyama, 1994).

However, very little is known about Earth's interior because of the limitations facing actual research. As such, the authors believe that integrating Earth science and planetary science as a new paradigm is surely the new research direction that could help us understand our Earth and, thus, terrestrial planets.

1.2 Nuclear Planetary Science

Planetary science, or planetology, is the science that focuses on how planetary systems and their components were formed, how they evolved, and how they will continue to evolve in the future. Our understanding of Earth and the planets has drastically evolved during the last few decades due to human advances in space exploration. Planetary bodies in our solar system are found to be extraordinarily diverse. Planetary science can be divided into two major classes, the study of the formation of the solar system through the initial conditions (or the primitive system) and the evolution of differentiated objects.

Planetary science involves a wide range of disciplines, skills, and expertise, including many of the natural sciences (e.g. astronomy, astrophysics, space plasma physics, atmospheric sciences, geology, geophysics, geochemistry) and basic disciplines (e.g. physical chemistry, hydrodynamics, material science, and scientific instrumentation). Successfully advancing the research, to a large extent, depends on how effectively these various disciplines can converge and cooperate to persue common questions. Planetary science can be divided into two major branches: one dealing with the study of the primitive solar system, and the other dealing with the evolution of differentiated bodies.

The Moon, Mars, and some asteroids are neighboring celestial bodies to our planet, Earth. Determining the distribution of major elements (Fe, Ti, Ca, Si, Mg, Al, O, etc.) and natural radioactive elements (K, Th, and U) in the surface materials of celestial bodies like the Moon, Mars, and asteroides, with no or thin atmosphere(s), as well as the distribution of volatile elements such as ice (formed from water) at polar regions, provides important clues to the conditions during the formation and evolution of their celestial bodies (Feldman *et al.*, 1998; Hiesinger and Head, 2006; Mitrofanov *et al.*, 2010; Pieters *et al.*, 2001; Warren, 1985; Warren and Wasson, 1979; Wieczorek and Zuber, 2004; Wilhelms *et al.*, 1987). The spectroscopy of neutrons, gamma rays, and X-rays is a powerful method for remotely measuring the absolute chemical abundances on their surface bodies (Adler and Trombka, 1977; Bielefeld *et al.*, 1976; Boynton *et al.*, 2004; d'Uston *et al.*, 2005; Feldman *et al.*, 1999; Hasebe *et al.*, 2008, 2009; Kim and Hasebe, 2012).

Surface materials of those celestial bodies are always exposed to cosmic rays (CRs). Because of nuclear interactions of CRs with the celestial bodies, neutrons are almost constantly produced. These neutrons leave the interaction sites and further hit other atoms in turn. They can excite atomic nuclei through inelastic scatterings. Such excited atoms emit gamma-rays. A part of the neutrons and gamma-rays produced in the materials leak from their surface, enabling nuclear spectrometers to decipher and reveal which elements are present. CR-excited major elements, along with naturally radioactive elements, can be measured by nuclear instruments in orbit (Reedy, 1978; Reedy and Trombka, 1973; Yamashita *et al.*, 2008).

Several first pioneering measurements of nuclear emissions from the Moon and Mars were carried out by J. Arnold and J. Trombka with colleagues more than 40 years ago, and by W.V. Boynton, C. d'Uston, N. Hasebe, D.J. Lawlence, T.H. Prettyman, I. Mitrofanov, J. Chan, and J.O. Goldsten with their colleagues about 10–15 years ago.

Today, planetary exploration missions conducted by several countries extend well beyond Earth into deep space. Japan has been participating in some rather conservative space exploration aimed at increasing our understanding of the solar system, whereas the United States and Europe have been engaging in several exciting and ambitious space explorations aimed at increasing our understanding of distant planets and their systems. Several countries have sent scientific observation missions to terrestrial planets: for example, the Lunar Prospector, Selene (Kaguya), Chang'E-1, -2, and -3, Chandrayaan-1 to the Moon, the Mars Odyssey, Smart-1 to the

Moon, Bepi Columbo to Mercury, and a few others to small bodies (Nears to Eros, Dawn to the Vesta and Ceres, Rosetta).

Nuclear emission from those celestial bodies is measured by remote nuclear sensing by space probes orbiting these bodies. Historically, Apollo and Luna have pioneered these investigations of the Moon in the 1960s (Metzger *et al.*, 1973; Vinogradov *et al.*, 1966). The Lunar Prospector mission in the late 1990s and Mars Observer in the early 2000s have performed global orbital imaging of the lunar and Martian emissions of gamma-rays and neutrons, which provided the first maps of the distribution of radioisotopes and soil constituting major elements over the entire lunar and Martian surfaces, respectively (Boynton *et al.*, 2002; Feldman *et al.*, 1998). Another important finding of these missions was a discovery of water ice with very high water content in the shallow subsurfaces of the Moon and Mars (Boynton *et al.*, 2002; Feldman *et al.*, 2001, 2002a, b; Mitrofanov *et al.*, 2002). In the late 2000s, Selene (Kaguya) (Kato *et al.*, 2008), Chang'E-1 (Chin *et al.*, 2007), and Chandrayaan-1 (Goswami *and Annadurai*, 2008) were launched as lunar missions after the Apollo, Luna, and Lunar Prospector missions.

X-ray and gamma-ray spectrometers (GRSs) were installed on these spacecraft. The scientific data from the Mars Odyssey and Selene (Kaguya) GRSs with high-purity Ge detectors have demonstrated the necessity of high spectral resolution, to reliably map key elements over the surface. However, the spatial accuracy of these maps are still limited by poor spatial resolution with the scale of an orbital altitude (Boynton *et al.*, 2004; Hasebe *et al.*, 2008, 2009; Kobayashi *et al.*, 2010, 2012). Another planetary mission that employed an HPGe in the GRS is the MESSENGER orbiting Mercury (Goldsten *et al.*, 2007).

The data from the GRS suite from the NASA Mars Odyssey have successfully demonstrated the research power of nuclear methods for Mars exploration: Actually, the data from GRS have completely changed the previous paradigm of Mars' cryosphere and hydrology. At present, several missions with nuclear instruments are either already making observations (NASA's Dawn: The Ceres after finishing the Vesta observation; NASA's Mars Science Laboratory — MSL for Mars) or are at the development stage (ESA's BepiColombo for Mercury, the Russian Phobos-Grunt for Phobos; the Korean Pathfinder Lunar Observatory).

The scientific study of planets, moons, and planetary systems, in particular, the study of the processes of their formation and evolution, is called planetology. Similarly, the study of the geology of the Moon falls under the more general field of lunar science, i.e., selenology. Nuclear planetology and selenology apply the nuclear radiation detection methodology to the geosciences, geochemistry, and planetology. In other words, we measure X-rays, gamma-rays, and neutrons emitted from planetary bodies with thin or no atmosphere(s). Here, we include not only gamma-ray and neutron spectroscopy but also X-ray spectroscopy to obtain the chemical information of the planetary bodies in the fields of nuclear planetology and selenology. Hereafter, we call nuclear planetology/selenology simply nuclear planetary science.

This method can be divided into two classes: remote sensing and *in situ* measurements. Remote nuclear planetology studies the planet by passively measuring GCR-induced and natural radioactive isotopes from the orbiter and/or through a fly-by mission (Boynton *et al.*, 2004; Hasebe *et al.*, 2008). *In situ* nuclear planetology is the study that passively measures GCR-induced and natural radioactive isotopes and/or actively combining neutron/X-ray source through landing/roving missions (Akkurt *et al.*, 2005; Goswami *et al.*, 2005; Hasebe *et al.*, 2011; Klingelhoefer *et al.*, 2007; Mitrofanov *et al.*, 2009; Parsons *et al.*, 2011; Trombka *et al.*, 2001).

Nuclear planetary science provides the unique and fruitful data of elements to study planets and small bodies of the solar system, which has not been possible with any other methods of investigation. The following are features of space experiments with nuclear instruments onboard orbiters and lander/rovers (Mitrofanov *et al.*, 2009): (1) the determination of the concentration of major elements and natural radioisotopes in the material of different planets and celestial bodies in the solar system to understand their origin and evolution; (2) the study of the distribution of volatile materials such as water H_2O, CO_2, and SO_2 in the subsurface of Mars, the Moon, and Mercury at some particular regions on their surfaces; and (3) the investigation of radiation conditions in the interplanetary space and on the surface of planets to determine potential radiation hazards during long-duration space flights and long-duration stays on the Moon and Mars.

The developments in nuclear physics and computational simulation and data analysis were introduced to nuclear geophysics. Today, these techniques developed in nuclear geophysics are also used in planetary science and astrophysics. Nuclear planetary science is the key to future *In-Situ* Resource Utilization (ISRU). ISRU enables and significantly reduces the mass and cost to produce materials and the risk of near-term and far-term explorations (Schrunk *et al.*, 2008; Wieczorek *et al.*, 2006). To search for important surface elements on the subsurface of the Moon, Mars, and asteroids, a neutron generator is an alternative source of GCRs and the active method that combines X-ray, gamma-ray, and neutron spectroscopy. Neutron generators can be used in future landing mission on the celestial bodies (Akkurt *et al.*, 2005; Mitrofanov *et al.*, 2009; Parsons *et al.*, 2011).

The main direction of further development of this method in nuclear remote sensing is thought to be the improvement of spatial resolution in the measurements from the orbit; to enable the identification of features and particular landscape objects on the surface being studied, in images of nuclear emissions. Spatial resolution, in the case of orbiting missions, is necessary for measurements in a scale of about 10–30 km.

In remote nuclear planetology, CRs play an important role in a celestial body's emission of gamma-rays and neutrons. Instead of CRs as incident projectiles, the active method of X-ray or neutron irradiation using a radiation generator enables us to measure major elements and the content of water ice/carbon dioxide in the polar soil for a short measuring interval (Akkurt *et al.*, 2005; Goswami *et al.*, 2005; Hasebe *et al.*, 2010; Klingelhoefer *et al.*, 2007; Mitrofanov *et al.*, 2009; Parsons *et al.*, 2011; Trombka *et al.*, 2001). Moreover, the instrumental concept should be based on increased spatial resolution for the surface mapping of X-ray, gamma-ray, and neutron emissions, with high spectral resolution and high sensitivity along the trace of the rover within about 1 m.

Before Apollo and Luna missions, our understanding of the Moon was a subject of almost unlimited speculation. Studies of both lunar samples and remote sensing data obtained by programs gave us the broad outline of the nature and geologic and geochemical history of the Moon. Now we know that the Moon is made of rocky material that has been variously melted, erupted through volcanic activity, and crushed by meteorite impacts. The Moon possesses a thick crust, a fairly uniform lithosphere, and an unconfirmed small iron core at the bottom of the asthenosphere. Some rocks hint at the possibility of an ancient magnetic field. The regolith was produced by innumerable meteorite impacts through geologic time.

However, knowledge and many beliefs formed after the Apollo mission are now being questioned on the basis of global data obtained by Clementine and the Lunar Prospector. In the 2000s, nuclear spectrometers on the Lunar Prospector, Selene (Kaguya), Chang'E, the Lunar Reconnaissance Orbiter, Mars Odyssey, Dawn, and MESSENGER provided precious global data and new findings. At present, data obtained from the Moon, Mars, Mercury, and asteroids in the late 1990s and 2000s are being integrated with new and old lunar sample data; giving us new ideas about the nature of the Moon.

Chapter 2

Principle of Nuclear Spectroscopy

Gamma rays, X-rays, alpha particles, and neutrons are emitted from various sources. We can obtain the source information of these photons and particles by detecting the photons and particles.

A gamma ray is an energetic photon mainly emitted when a nucleus either moves from a higher excited state to a lower excited state or the ground state; or when a positron combines with an electron and thus annihilate one another. Since the energy levels of the excited states are specific to the nuclei that produced them, the energy of each gamma ray allows scientists to identify the source nucleus.

An X-ray is a photon with energy higher than an ultraviolet photon. An X-ray is emitted when an orbital electron transits from the higher excited state to either the lower excited state or the ground state. Since the energy levels of the excited states are specific to each element, the energy of an X-ray allows scientists to identify the source element. An X-ray is also emitted when a charged particle is decelerated in the electric field of a nucleus, and is called 'bremsstrahlung'. The energy of a bremsstrahlung X-ray is distributed continuously. In our solar system, solar X-rays excite atoms on a celestial body with no or little atmosphere, and the excited atoms emit a type of X-ray called fluorescent X-rays.

Some nuclei of the uranium and thorium series, such as ^{222}Rn and ^{210}Po, emit alpha particles. The energy of an alpha particle is characteristic of the source nucleus; however, when the alpha particle passes through matter, even a very thin layer, energy is lost. Therefore, the energy of alpha particles identifies the source nuclei only in cases where no substance stands between the source nuclei and the detector.

Neutrons are emitted by nuclear interactions. These neutrons interact with nuclei through elastic and/or inelastic collisions and emit protons or neutrons losing the energy. Fast neutrons lose energy efficiently through elastic collisions with light nuclei. So the energy distribution of neutrons at a site reflects the elemental composition of the site, though the neutron energy does not uniquely correspond to the source interaction. Moreover, nuclei that are excited by neutrons may emit gamma rays. Losing their energy through collisions with atoms in the material, some neutrons become thermal neutrons, which may be efficiently captured by some nuclei. These nuclei may emit gamma rays. So, the energy and the flux of neutrons are necessary to determine the elemental abundance from the intensity of neutron-induced gamma rays.

A lander or rover could be put on a celestial body to collect gamma ray, X-ray, and/or neutron data. Such a probe may carry the source of X-rays, alpha particles, or neutrons to excite atoms and/or nuclei and found there, and — like what is done during petroleum exploration on the earth — may observe fluorescent X-rays, neutrons, or gamma rays on the body.

2.1 Natural Source of Planetary Gamma Rays

Gamma rays are emitted from excited nuclei, which are in nature caused by the decay of natural radioactive nuclei or by nuclear interactions of energetic charged particles of cosmic rays.

2.1.1 *Natural radionuclides emitting gamma rays and alpha particles*

Radioactive nuclei with long half-lives, such as ^{40}K, ^{238}U, and ^{232}Th have been present since the formation of the solar system about 4.5 G years ago. These radioactive nuclei decay and produce excited nuclei. ^{40}K decays into ^{40}Ca or ^{40}Ar with a half-life of 1.3 G years. In the decay into ^{40}Ar, gamma rays with energy of 1.46 MeV is emitted. ^{238}U decays into

^{234}Th, with a half-life of about 4.5 G years; during this decay, alpha particles are emitted. ^{234}Th is also unstable and decays to ^{234}Pa, with a half-life of about 24 days; electrons are emitted during they decay. Figure 2.1 shows that ^{238}U transforms into a stable isotope of ^{206}Pb after nine alpha decays and six beta decays. The chain of decay series branches off and merges until the final ^{206}Pb and the 18 unstable isotopes — including the first ^{238}U (called uranium series, or radium series, isotopes) are reached. ^{232}Th also decays into ^{228}Ra with a half-life of about 14 G years, and emits alpha particles during the decay. The chain of decay terminates at the stable isotope of ^{208}Pb. Thus ^{232}Th becomes stable ^{208}Pb after six alpha decays and four beta decays through 10 unstable isotopes. The first ^{232}Th, the final ^{208}Pb, and 10 unstable isotopes are called thorium series isotopes. Some of the unstable nuclei in the uranium series and thorium series emit gamma rays. Besides the above isotopes, ^{235}U and ^{237}Np initiate other decay series that are less abundant due to the shorter half-lives of ^{235}U and ^{237}Np in comparison to the half-lives of ^{238}U or ^{232}Th. There are some independent natural radioactive isotopes such as ^{138}La and ^{176}Lu.

If a mother nuclide has a life sufficiently longer than the daughter, granddaughter, and descendant nuclide, the number of disintegrations of the daughter, granddaughter, and descendant nuclide finally equals to that of mother nuclide per unit time. In the case that the mother and descendant nuclides are in the same state, the number N_i of the respective nuclide by the half-life τ_i of the nuclide (N_i/τ_i) equals to N/τ for all descendant and mother nuclides. This state is called the radioactive equilibrium. The difference between N_i/τ_i indicates the past event disturbing the equilibrium and provides chronological information in some cases.

In a decay of a radioactive element, a certain amount of energy is released. The radioactive element serves as a heat source of the celestial body. Therefore, the abundance of natural radioactive elements is important in the thermal history, differentiation, and evolution of the celestial body.

Reedy (1978) estimated the flux of gamma rays emitted by the natural radioactive nuclides on the surface of Earth's Moon (shown in Table 2.1) by assuming the elemental composition of the lunar surface.

$_{92}$U $_{91}$Pa $_{90}$Th $_{89}$Ac $_{88}$Ra $_{87}$Fr $_{86}$Rn $_{85}$At $_{84}$Po $_{83}$Bi $_{82}$Pb $_{81}$Tl $_{80}$Hg

^{238}U 4×10⁹y → α → ^{234}Th 24 d ↙ β

^{234}Pa 1 m ↙ β

^{234}U 2×10⁵y → α → ^{230}Th 8×10⁴y → α → ^{226}Ra 1600 y → α → ^{222}Rn 4 d → α → ^{218}Po 3 m → α → ^{214}Pb 27 m

U–series

^{218}At 2 s → α → ^{214}Bi 20 m → α → ^{210}Tl 1 m ↙ β ↙ β ↙ β

^{218}Rn 0.035 s → α → ^{214}Po 160 μs → α → ^{210}Pb 22 y → α → ^{206}Hg 8 m ↙ β ↙ β

^{210}Bi 5 d → α → ^{206}Tl 4 m ↙ β ↙ β

^{210}Po 140 d → α → ^{206}Pb stable

^{232}Th 1x10¹⁰y → α → ^{228}Ra 6 y ↙ β

^{228}Ac 6 h ↙ β

^{228}Th 2 y → α → ^{224}Ra 4 d → α → ^{220}Rn 60 s → α → ^{216}Po 0.15 s → α → ^{212}Pb 11 h

Th–series

^{212}Bi 61 m → α → ^{208}Tl 3 m ↙ β ↙ β

^{212}Po 0.3 μs → α → ^{208}Pb stable

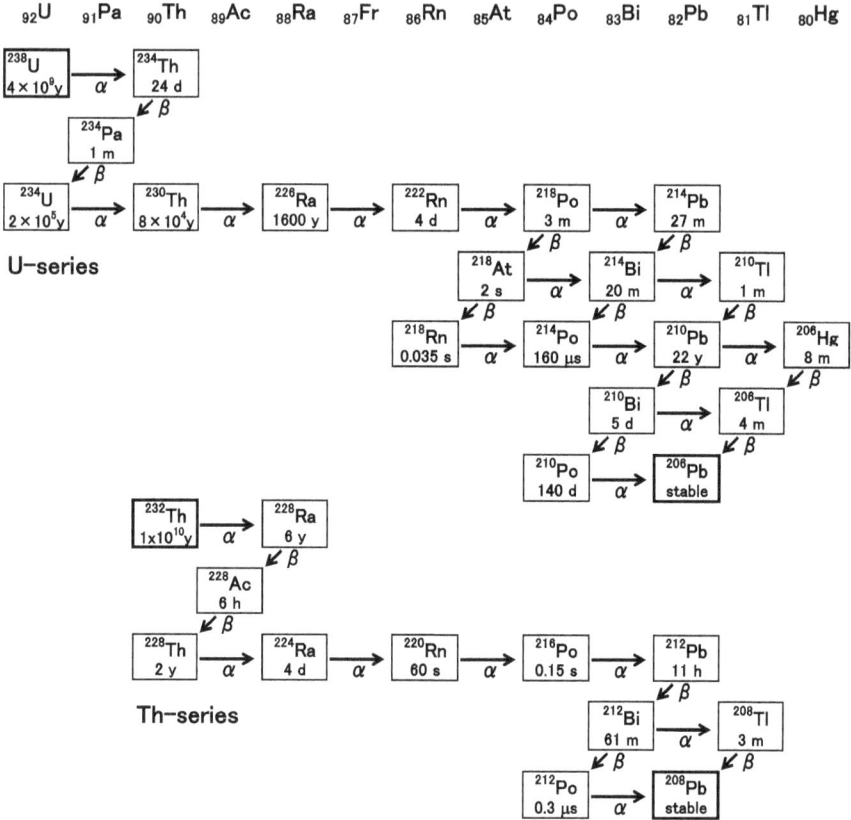

Figure 2.1: U-series and Th-series decay chains. Some minor branches, e. g. in ^{234}Pa, are omitted.

In U-series isotopes (Table 2.1), 13 isotopes emit alpha particles, and in Th-series, 7 isotopes emit alpha particles. These alpha particles have energies specific to the respective isotopes. For example, ^{210}Po emits 5.305 MeV alpha particles, and ^{222}Rn emits 5.490 MeV alpha particles. So, by detecting the initial energy of alpha particles, we can identify the mother isotope. Unfortunately, an alpha particle loses energy in a short distance. For example, the range of a 5.0 McV alpha particle is 6.63 mg/cm² or 25 μm in aluminum. This means that we can identify the mother isotope by measuring the energy of an alpha particle at the orbit altitude only in scenarios where the celestial body has no atmosphere and the mother isotope is absolutely on the surface of the celestial body. In the case of our Moon, U-series and Th-series isotopes are abundant in its rock and soil. These isotopes are expected to appear on the lunar surface through the boundary of the lunar crust or volcano because some radon isotopes are included in the decay chain of these series and the radon gases may have large diffusion coefficients. It is expected that detecting alpha particles will provide the key to understanding the underground structure of the Moon. In the Apollo 15 and 16 experiments, alpha particles emitted from ^{210}Po and ^{222}Rn were detected (Adler *et al.*, 1972a; 1972b).

Figure 2.2 shows the gamma-ray spectrum obtained with a small Ge detector for the soil sampled near Tokyo. The soil sample contained naturally radioactive elements listed in Table 2.2. Figure 2.2 illustrates many peaks of gamma rays emitted from naturally radioactive elements including ^{40}K, ^{208}Tl in the Th-series, and ^{214}Bi in the U-series.

2.1.2 *Cosmic ray-produced gamma rays and neutrons*

On a celestial body with no or little atmosphere, cosmic ray particles such as fast protons and ^{4}He collide with nuclei on the surface of the body and produce excited nuclei, protons, and neutrons. The secondary protons and neutrons may further collide with nuclei and produce excited nuclei, protons, and neutrons as long as the energy is high enough to interact with other nuclei. The cosmic rays include galactic cosmic rays with high energy particles and solar cosmic rays with lower energy particles. The neutrons may collide either elastically or inelastically with nuclei and produce excited nuclei. The excited nuclei emit gamma rays. Losing the energy by elastic or inelastic collisions with nuclei, the neutron may be absorbed by a nucleus, which may emit gamma rays. Therefore, the irradiation of cosmic rays induces the emission of gamma rays. The abundant elements significantly interact with such neutrons and emit gamma rays. So, these gamma rays significantly reflect the abundance of major elements.

Table 2.1:   Flux of gamma rays produced by the decay of natural radionuclide at the lunar elemental abundances.

Element	Nuclide	Energy (MeV)	Yield	Flux (photons/cm² min)
K	^{40}K	1.4608	0.1048	2.352
La	^{138}La	1.4359	0.671	0.00247
	^{138}La	0.7887	0.329	0.00090
Lu	^{176}Lu	0.3069	0.94	0.00756
	^{176}Lu	0.2018	0.85	0.00573
Th	^{208}Tl	2.6146	0.360	2.193
	^{228}Ac	1.6304	0.019	0.0919
	^{212}Bi	1.6205	0.016	0.0771
	^{228}Ac	1.5879	0.037	0.177
	^{228}Ac	1.4958	0.010	0.0464
	^{228}Ac	1.4592	0.010	0.0458
	^{228}Ac	0.9689	0.175	0.656
	^{228}Ac	0.9646	0.054	0.202
	^{228}Ac	0.9111	0.290	1.054
	^{208}Tl	0.8605	0.045	0.159
	^{228}Ac	0.8402	0.010	0.0349
	^{228}Ac	0.8356	0.018	0.0627
	^{228}Ac	0.7948	0.048	0.163
	^{212}Bi	0.7854	0.010	0.0338
	^{228}Ac	0.7721	0.016	0.0537
	^{228}Ac	0.7552	0.011	0.0366
	^{212}Bi	0.7271	0.070	0.229
	^{208}Tl	0.5831	0.307	0.916
	^{228}Ac	0.5623	0.010	0.0294
	^{208}Tl	0.5107	0.083	0.234
	^{228}Ac	0.4630	0.046	0.125
	^{228}Ac	0.4094	0.022	0.0566
	^{228}Ac	0.3384	0.120	0.285

(*Continued*)

Table 2.1:   (*Continued*)

Element	Nuclide	Energy (MeV)	Yield	Flux (photons/cm² min)
	^{228}Ac	0.3280	0.034	0.0797
	^{212}Pb	0.3000	0.031	0.0700
	^{208}Tl	0.2774	0.024	0.0524
	^{228}Ac	0.2703	0.038	0.0821
	^{224}Ra	0.2410	0.038	0.0783
	^{212}Pb	0.2386	0.47	0.964
	^{228}Ac	0.2094	0.045	0.0874
U	^{214}Bi	2.4477	0.016	0.0753
	^{214}Bi	2.2041	0.050	0.224
	^{214}Bi	2.1185	0.12	0.0526
	^{214}Bi	1.8474	0.021	0.0861
	^{214}Bi	1.7645	0.159	0.637
	^{214}Bi	1.7296	0.031	0.123
	^{214}Bi	1.6613	0.012	0.0467
	^{214}Bi	1.5092	0.022	0.0817
	^{214}Bi	1.4080	0.025	0.0897
	^{214}Bi	1.4015	0.014	0.0501
	^{214}Bi	1.3777	0.040	0.142
	^{214}Bi	1.2810	0.015	0.0514
	^{214}Bi	1.2381	0.059	0.199
	^{214}Bi	1.1552	0.017	0.0554
	^{214}Bi	1.1203	0.150	0.481
	^{214}Bi	0.9341	0.032	0.0939
	^{214}Bi	0.8062	0.012	0.0328
	^{214}Pb	0.7859	0.011	0.0297
	^{214}Bi	0.7684	0.049	0.131
	^{214}Bi	0.6655	0.016	0.0403

(*Continued*)

Table 2.1:   (*Continued*)

Element	Nuclide	Energy (MeV)	Yield	Flux (photons/cm² min)
	^{214}Bi	0.6093	0.461	1.118
	^{214}Pb	0.3519	0.371	0.714
	^{214}Pb	0.2952	0.192	0.343
	^{214}Pb	0.2419	0.075	0.123
	^{226}Ra	0.1860	0.055	0.0810
	^{235}U	0.1857	0.54	0.0366

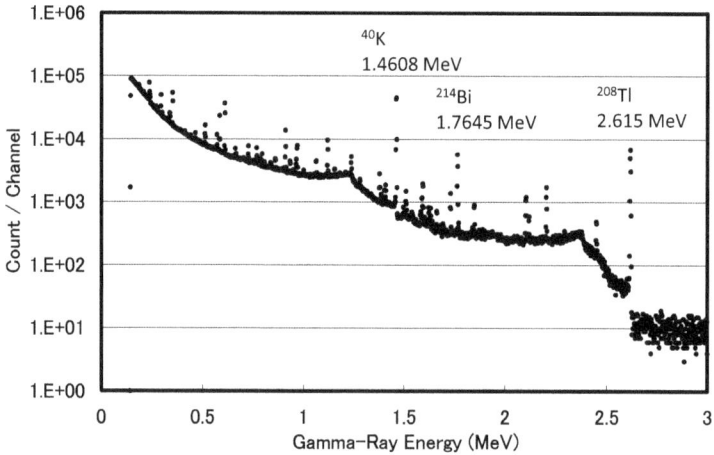

Figure 2.2:   Gamma-ray spectrum obtained with a Ge detector for the soil sampled near Tokyo. The relative efficiency of the Ge detector was 12%.

The flux of galactic cosmic rays as a function of the energy depends on the solar activity with a cycle of about 11 years. When solar activity is high, the solar magnetic field is strong and efficiently blocks the galactic cosmic rays with lower energy. Figure 2.3 shows the energy spectra of cosmic ray protons near Earth in three solar conditions (Yamashita *et al.*, 2008). As seen in the figure, protons with energy less than 1 GeV are appreciably affected by the solar activity.

Table 2.2: Energies of major gamma-ray lines expected from the elements on Moon (Yamashita *et al.*, 2008).

Elements	Gamma-ray energy [MeV]				
H	2.223^c				
O	6.129^n	4.438^n	5.269^n		
Mg	1.369^n	1.809^n			
Al	1.014^n	2.211^n	0.844^n	7.724^c	
Si	1.779^n	3.539^c	4.934^c		
S	2.230^n	5.421^c			
Ca	1.943^c	3.737^n	6.420^c		
Ti	6.760^c	6.418^c	1.382^c	0.342^c	0.984^n
Fe	7.631^c	7.646^c	0.847^n		
K	1.461^s				
Th	2.615^s	0.911^s	0.239^s	0.583^s	
U	0.609^s	1.764^s	0.296^s		

The superscripts for each energy indicate the following reactions that cause gamma-ray emission; c: capture of neutrons, n: nonelastic scattering, s: spontaneous decay of naturally radioactive nuclides.

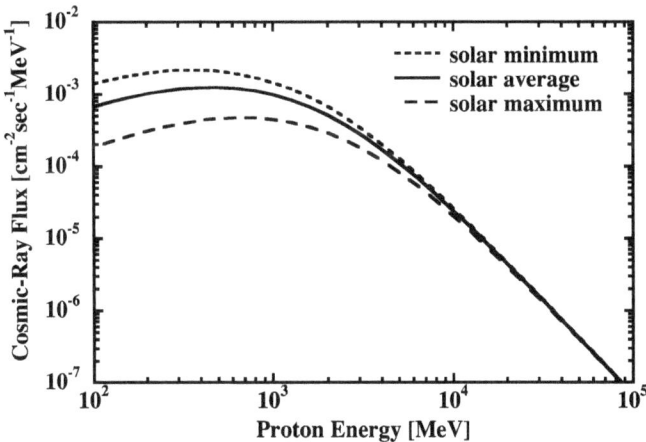

Figure 2.3: Energy spectra of galactic cosmic ray protons at 1 AU under different solar conditions: minimum, average, and maximum (Yamashita *et al.*, 2008).

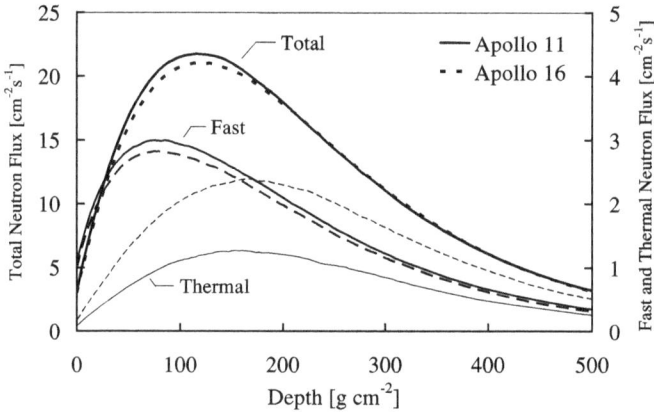

Figure 2.4:   Calculated neutron flux as a function of depth from the lunar surface (Yamashita *et al.*, 2008).

Incident cosmic rays interact with atomic nuclei and produce neutrons on the surface of Moon. These neutrons collide with nuclei elastically or inelastically, produce further neutrons, lose the energy, and are absorbed by nuclei or escape from the surface of Moon. The production of neutrons is affected by the composition of the surface material. Figure 2.4 shows the flux of neutrons as a function of the depth from the lunar surface for the two cases; one is the composition near the site of Apollo 11 and the other is the composition near the site of Apollo 16 (Yamashita *et al.*, 2008). As seen in the figure, fast neutron density is highest at depths below 100 g/cm^2 and the density of slow neutrons is highest at depths near 200 g/cm^2. Neutrons efficiently lose energy during elastic collisions with light nuclei; they do so most efficiently with hydrogen nuclei. So the production and the modulation of neutrons are affected by the composition of the surface. The flux and energy spectra of the neutrons reflect the composition of the surface. The energy spectra of the escaped neutrons are shown in Fig. 2.5 (Yamashita *et al.*, 2008) for the model composition of Apollo 11, Apollo 16, and ferroan anorthosite (FAN). Yamashita *et al.* (2008)'s estimates of the energies of gamma-ray lincs expected from the Moon are listed in Table 2.2.

Gamma rays are emitted from the excited nuclei as a result of inelastic scattering with fast neutrons or through the absorption of slow neutrons.

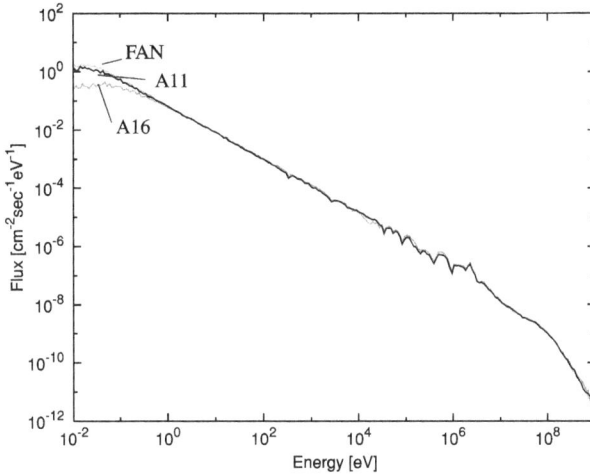

Figure 2.5:   Energy spectra of leakage neutrons from the lunar surface (Yamashita *et al.*, 2008).

Some of the gamma rays from the surface of the Moon leak into space. As described earlier, the neutrons and gamma rays leaked reflect the elemental composition of the lunar surface and may be detected from spacecraft, as demonstrated by Apollo projects, Lunar Prospector, KAGUYA, and so on. Feldman *et al.* (1998) reported evidence of water ice at the lunar poles through the energy spectra of neutrons obtained by Lunar Prospector.

2.1.3 *Spatial resolution in gamma-ray observation*

Gamma rays are more penetrative than other photons with lower energies. Using a massive collimator, one may restrict a gamma-ray spectrometer's field of view and improve the spatial resolution. In the remote sensing of gamma rays in a spacecraft's orbit, a massive collimator is not practical, and gamma-ray detectors in general have nearly isotropic sensitivity. Moreover, gamma rays approach such a detector from all areas of the celestial body in sight. For example, in a lunar orbit at an altitude of 100 km, the horizon of Moon is 598 km apart from the satellite. Here, we assumed that the shape of Moon is a sphere with a

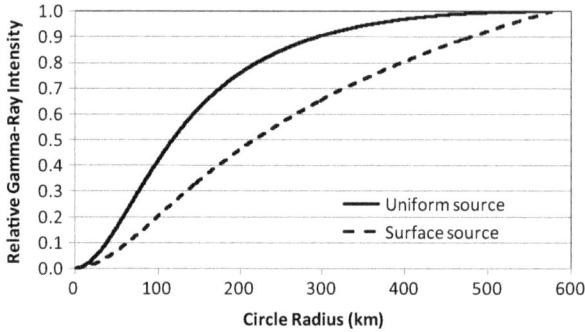

Figure 2.6:   The ratio of detected gamma rays emitted from the inside of a circle with radius *r* for uniform and surface source distributions. The radius is measured along the surface of the Moon from the nadir of the satellite at an altitude of 100 km over the Moon.

radius of 1738 km. This means that the horizon is about 576 km from the nadir along the Moon's surface. On the extreme assumption that gamma rays are isotropically emitted from the exact lunar surface, that is, that the surface material of the Moon does not attenuate the gamma ray, 50% of the detected gamma rays are emitted in a circle with a radius of about 220 km along the surface of Moon, as shown by the "surface source" curve in Fig. 2.6.

More practically, one may assume that the gamma-ray source is distributed uniformly along the depth from the surface. This is a better assumption for gamma rays emitted from naturally radioactive isotopes. In this case, 50% of the gamma rays hitting the detector originate in a circle with a radius of about 120 km, as shown by the "uniform source" curve in Fig. 2.6. The difference in the radii is due to the collimation of gamma rays by the surface material. For example, the ratio of gamma ray intensity I_β from a point from which the zenith angle of the satellite is β to the gamma ray intensity I_N from the nadir is $I_\beta/I_N = \cos\beta$ for uniform distribution of gamma-ray source. In this case, one may regard the spatial resolution of the gamma-ray spectrometer (GRS) to approximately be a circle with a radius of around 120 km at an orbital altitude of 100 km.

Cosmic rays produce neutrons that are not uniformly distributed along the depth, as shown in Fig. 2.4. The neutron flux is maximum near the depth of about 0.5–1 m for fast neutrons and near the depth of about 1–2 m for thermal neutrons, depending on the elemental abundance near the surface. So, the effect of the collimation by the lunar material is appreciable for gamma rays from the neutron interactions more than that by naturally-occurring radioactive isotopes. Reedy *et al.* (1973) calculated the space resolution for the three sources of gamma rays: uniformly distributed source, (n, γ) reaction, and (n, Xγ) reaction for the estimation of lunar chemical composition. Thermal neutrons mainly cause (n, γ) reactions, and therefore the collimation effect for gamma rays by an (n, γ) reaction is the most significant (Reedy *et al.*, 1973).

The altitude of the satellite affects the spatial resolution. By descending the orbit spatial resolution in gamma-ray observation is improved. Figure 2.7 shows the radius, $R_{1/2}$, of a circle that emits half of the gamma rays hitting the detector, as a function of an orbit altitude H over the Moon for uniform source distribution along the depth. The radius $R_{1/2}$ of 120 km at $H = 100$ km reduces to 65 km at $H = 50$ km. For neutron-induced gamma rays, the radius $R_{1/2}$ is smaller. Lunar Prospector and KAGUYA made gamma-ray observations at a descended altitude around 50 km.

For a planet with a thick atmosphere, gamma rays from the surface are absorbed by its atmosphere, thereby making observations from a spacecraft in orbit impractical. For example, Earth's atmosphere has a thickness of 10^3 g/cm^2 and attenuates 1 MeV gamma rays from the ground to less

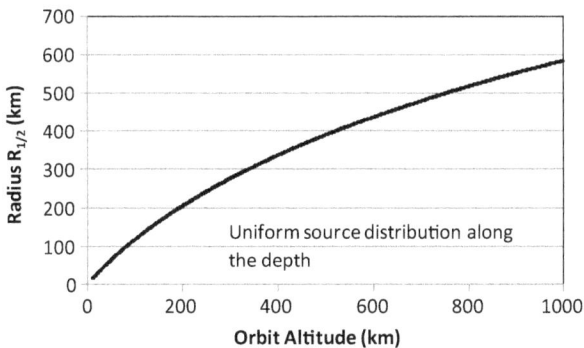

Figure 2.7:   Radius $R_{1/2}$ of a circle from which half of the measured gamma rays originate as a function of the spacecraft altitude H. The source of the gamma ray is assumed to be uniform along the depth.

than 10^{-28}. For a planet with a thin atmosphere, such as Mars, gamma-ray observation can be carried out from a spacecraft in orbit. Such a thin atmosphere acts not only as an attenuator but also as a collimator for gamma rays.

2.1.4 *X-rays induced by solar X-rays*

An X-ray is emitted from an excited atom when the atom moves to lower excited states or to the ground state. An X-ray emitted from an atom excited by high-energy X-rays is called a fluorescent X-ray. Table 2.3 shows the energy of Kα X-rays from elements with atomic numbers from 4 through 36. As seen in the table, the energy of each X-ray is specific to the element and is called a characteristic X-ray. Therefore, the atomic number and the concentration of the element emitting the X-ray can be determined from the energy and the intensity of the X-ray.

In an orbital X-ray spectroscopy of a celestial body such as the Moon, only solar X-rays are potentially intense enough to excite the atoms on the body. However, the intensity of the solar X-rays depends on the condition of the Sun and are typically not stable in intensity and energy spectrum.

Careful consideration is needed to derive elemental abundance from X-ray spectroscopy. From the Apollo 15 and 16 explorations, the ratios of Al/Si atoms and Mg/Si atoms were obtained over part of a region near the equator of the Moon.

2.2 Active Sources of Neutrons and X-rays

In the case of missions involving a lander or rover on a celestial body, it is possible to analyze fluorescent X-rays by artificially exciting atoms on the surface. The analysis of neutrons and gamma rays is also possible by using an artificial source of neutrons.

2.2.1 *Active sources of neutrons*

Neutrons are emitted from light nuclei irradiated by alpha particles or by gamma rays. Interacting with alpha particles from ^{241}Am, ^{9}Be emits fast

Table 2.3:   Energy of Kα X-ray.

Atomic number	Element	Kα (keV)
4	Be	0.11
5	B	0.183
6	C	0.277
7	N	0.392
8	O	0.525
9	F	0.677
10	Ne	0.848
11	Na	1.041
12	Mg	1.253
13	Al	1.486
14	Si	1.739
15	P	2.013
16	S	2.307
17	Cl	2.621
18	Ar	2.957
19	K	3.312
20	Ca	3.69
21	Sc	4.088
22	Ti	4.508
23	V	4.949
24	Cr	5.411
25	Mn	5.894
26	Fe	6.398
27	Co	6.924
28	Ni	7.471
29	Cu	8.04
30	Zn	8.63
31	Ga	9.241
32	Ge	9.874
33	As	10.53
34	Se	11.207
35	Br	11.907
36	Kr	12.631

neutrons with average energy of 5.0 MeV. When irradiated by gamma rays from ^{24}Na, ^{9}Be emits fast neutrons with an average energy of 0.83 MeV. About 3% of ^{252}Cf undergo spontaneous fission and emit about 3.8 neutrons with an average energy of 2 MeV, per fission. About 97% of ^{252}Cf emit alpha particles with an average kinetic energy of about 6.2 MeV. Pyroelectric crystals have also recently been investigated as a source of neutrons (Naranjo *et al.*, 2005) as well as X-rays.

2.2.2 *Active sources of X-rays*

An atom may be excited by a photon or a charged particle and emit photons. The emitted photon is called an X-ray if the energy of the photon is high. X-rays are usually generated by an X-ray tube, where electrons are accelerated in an electric field through the application of a high voltage. The accelerated electrons are focused onto a target that is often made of a relatively heavy element. These electrons may generate bremsstrahlung X-rays with an energy distribution below the initial electron's energy to zero energy in the target. The X-rays produced may be characteristic of the atom in the target. The characteristic X-ray is dominant in the case where the target is made of a relatively light element such as those found in pyroelectric crystals. A pyroelectric crystal, e.g. lithium tantalate ($LiTaO_3$) or lithium niobate ($LiNbO_3$), with a thickness of several millimeters, can generate voltage as high as 100 kV when the temperature of the crystal is changed from room temperature to 100°C. Given the recent commercialization of compact and lightweight X-ray generators using pyroelectric crystals as their high-voltage source, such generators might soon be available for X-ray spectrometry in future planetary explorations (Kusano *et al.*, 2014; 2016).

Alpha particles can induce the characteristic X-rays from atoms on the surface of a celestial body. The Mars rover Opportunity carries an X-ray spectrometer and a ^{244}Cm source. The spectrometer detects characteristic X-rays from Mars, irradiated by alpha particles from the ^{244}Cm source. The ^{244}Cm source may be replaced with ^{55}Fe and ^{109}Cd sources. The radioactive isotope ^{55}Fe captures electrons and its daughter emits

5.9 keV X-rays. Radioactive ^{109}Cd captures electrons, and its daughter emits 88 keV gamma rays and 22 keV X-rays. X-rays from ^{55}Fe are effective in exciting elements slightly lighter than Fe, and those from ^{109}Cd are effective in detecting atoms slightly lighter than an Ag atom. These isotopes are used in X-ray spectrometry for detecting major elements in planetary exploration.

Chapter 3

Nuclear Spectrometer

Incidences of radiation such as gamma rays, neutrons, alpha particles, and X-rays may be detected as electronic signals by detectors suitable for measuring radiation and kinetic energy.

3.1 Gamma-Ray Detector

Gamma rays with energy E_γ transfer their energy to electrons through three interactions: (1) the photoelectric effect, (2) Compton scattering, and (3) pair creation. (1) In the photoelectric effect, a gamma ray vanishes, transfering its energy to an orbital electron of an atom in the material. The electron is liberated with the kinetic energy $E_e = E_\gamma - I$, where I is the ionization potential of the electron. The ionized atom emits a photon or photons corresponding to energy I. These photons are called X-rays, which may further ionize atoms. (2) In Compton scattering, a gamma ray elastically collides with an electron in the material and is scattered, where a part of the photon energy is transferred to the electron according to the scattering angle. The gamma ray survives, losing a part of the energy. (3) In pair creation, a gamma ray with energy larger than 1.022 MeV produces an electron–positron pair near the nuclei of the atom in the material, where the gamma ray vanishes. Losing the kinetic energy, the positron may cause pair annihilation in the material, where the positron combines with an electron in the material and the electron–positron pair annihilates by emitting two gamma rays with energy of 0.511 MeV.

In Compton scattering and pair creation, the energy of the initial gamma ray is shared by electron(s) and/or gamma ray(s) with lower energy. The secondary gamma rays may further cause these three interactions until gamma rays vanish or exit the material. Also, the liberated electrons may produce energetic photons or electrons. The initial gamma ray is identified in cases where the total energy is deposited in the material working as the detector.

The efficiency of photoelectric conversion by an atom with an atomic number Z is roughly proportional to $Z^n/E_g^{-3.5}$, where E_g is the energy of the gamma ray and the index n is 4–5, depending on E_g. This means that a large atomic number is essential to detect energetic gamma rays with photoelectric conversion. The density of the detector reflects the number density of electrons, so dense detectors effectively scatter gamma rays by Compton scattering. Pair creation efficiently occurs at strong electric fields near a heavy nucleus. Therefore, the gamma-ray detection efficiency of the detector with a given size increases with the atomic number and the density of the detector.

The probability (P) of a gamma ray surviving with the initial energy after passing through the material with a thickness t (g/cm^2) is given by

$$P = \exp(-\mu t),$$

where μ is the mass attenuation coefficient. The mass attenuation coefficient depends on the atomic number of the material and on energy of the gamma ray (E_g).

Gamma rays after Compton scattering and those created by pair creation may interact again, and these interactions may succeed in a cascade in the material according to the size of the material and energy of the gamma ray, until the gamma rays vanish in or exit from the material. In the case that all or a part of the gamma ray energy is given in the detector material, the corresponding signal is recorded and thus the incident gamma ray is detected. A part of the incident gamma ray gives the total energy to electron(s) in the material through a single photoelectric effect or multiple interactions, which can include the photoelectric effect, Compton scattering, and pair creation. In this case, the total energy of the released electron(s) equals the energy of the incident gamma ray. In gamma-ray spectrometry, we know the energy of gamma rays practically from gamma rays that transfer the total energy to the detector. The ratio of the number of gamma rays whose total or partial energy is absorbed in the detector material to the number of gamma rays entering the detector is called the intrinsic total detector efficiency. The ratio of the number of gamma rays detected to the number of gamma rays entering the detector is called the absolute total detector efficiency. The ratio of the number of gamma rays whose total energy is absorbed in the detector to the number of gamma rays whose partial or total energy is absorbed in the detector is called the photo-peak efficiency.

The probabilities of the aforementioned three interactions are high in detectors of a large size, with high density, and with an element of a larger atomic number. To obtain sufficient detection efficiency, the cross section and/or thickness of the detector should be large.

As described earlier, incident gamma rays produce energetic electrons as a result of interactions with materials. These electrons ionize or excite atoms along the track. Some of the electrons liberated in ionization may further ionize or excite other atoms depending on the energy. This process continues until all the electrons reduce their kinetic energy to a level below the excitation energy of atoms in the detector. In semiconductors, energetic electrons produce electron–hole pairs and excited states. The ionized electrons and ions, or holes, are collected by applying a sufficient electric field. In the absence of an electric field, electron–ion pairs or electron–hole pairs recombine and form excited atoms or excited states. These excited atoms and excited states may emit photons, which we may detect. From the collected charge or number of detected photons, we determine the energy deposited in the detector by the incident gamma ray.

3.1.1 *Semiconductor detectors*

Energetic electron(s) made by a gamma ray produce(s) electron–hole pairs and excited states in semiconductors. In cases where gamma rays give energy (E) in the detector, the average number (N_i) of electron–hole pairs produced is given by $N_i = E/\varepsilon$, where ε is the average energy to produce one electron–hole pair in the semiconductor. By collecting the charge of electrons and holes, N_i is obtained and the energy given by the gamma ray is known to be $E = \varepsilon N_i$. The statistical fluctuation of N_i is given by $(f \cdot N_i)^{1/2}$, where f is the Fano factor. This corresponds to the fluctuation in energy (ΔE), which is calculated as follows:

$$\Delta E = \varepsilon(f \cdot N_i)^{1/2} = (f \cdot \varepsilon \cdot E)^{1/2}$$

This is typically the main factor limiting the energy resolution in Ge semiconductors, as described next.

The major semiconductors used for gamma-ray detection are Ge, CdTe, HrI_2, and CZT. Silicon semiconductors are the most popular semiconductors and are often used to detect charged particles and X-rays. Their application to energetic gamma rays, however, is not popular due to the small atomic number of Si. Table 3.1 summarizes the properties of these semiconductors (Knoll, 1999).

Germanium semiconductors have a narrow band gap of 0.72 eV, which results in small ε value and a large leak current at room temperature. The small ε value contributes to the good energy resolution. But due to the large leak current at room temperature, germanium semiconductors should be operated at a cryogenic temperature to reduce the noise caused by the leak current. In germanium semiconductors, mobility is large and the mean free time is long for each electron and hole. This allows us to

Table 3.1:   Properties of semiconductors for X-and gamma ray detection.

Semiconductor (operating temperature)	Atomic number	Density (g/cm³)	Band gap (eV)	ε (eV)	Best energy resolution/ gammaray energy
Si (300 K)	14	2.33	1.12	3.61	400 eV/60 keV
Si (77 K)			1.16	3.76	550 eV/122 keV
Ge (77 K)	32	5.33	0.72	2.98	1.3 keV/1.33 MeV
					900 eV/662 keV
					400 eV/122 keV
CdTe (300 K)	48/52	6.02	1.52	4.43	3.5 keV/122 keV
					1.7 keV/60 keV
HgI₂ (300 K)	80/53	6.36	2.13	4.3	5.96 keV/662 keV
					1.7 keV/60 keV
CZT (300 K)	48/30/52	6	1.64	5.0	11.6 keV/662 keV

measure energetic gamma rays with good energy resolution and with high efficiency by employing large germanium semiconductors.

In addition to the description given in Table 3.1, diamond semiconductors have large mobility for electron and hole and might be applicable to detect charged particles in a wide range of temperature; however, the small atomic number of carbon causes very low detection efficiency for energetic gamma rays.

As shown in Table 3.1, compound semiconductors have wider band gaps and can be operated at room temperature without an appreciable leak current, though the energy resolution would be less than that of germanium conductors. The mobility (μ_e and μ_H) and mean life time (τ_e and τ_H) of the electron and hole are essential for the semiconductors used as gamma-ray detectors. The products of $\mu_e\tau_e$ and $\mu_H\tau_H$ give the "mean free path" of the charge carrier in the semiconductor. Most compound semiconductors have low μ_e, μ_H and short τ_e, τ_H, in particular, low μ_H and short τ_H. Due to the resulting difficulty in charge collection, the size of the compound semiconductors is limited, and therefore the detection efficiency is low — particularly for energetic gamma rays. As seen in Table 3.1, the compound semiconductors are mainly applied for the detection of low-energy gamma rays. The energy resolution in germanium detectors, so far, is the best among these semiconductors. Chandrayaan-1 employed the array of CZT semiconductor detectors for the measurement of X-rays and low-energy gamma rays (Goswami *et al.*, 2005). The array structure is useful for increasing the detector's detection efficiency of energetic photons.

3.1.2 *Germanium semiconductor detectors*

Germanium detectors are operated at a cryogenic temperature, typically below 90 K, with some cryogenic system. A popular cryogenic system is that with liquid nitrogen in the ground operation. For germanium detectors used in planetary observation, a mechanical cooler or radiative cooler is usually employed. The Mars Observer and Mars Odyssey each carried a germanium gamma-ray spectrometer with a radiative cooling system (Boynton *et al.*, 2004). The radiative cooling system has some merits; namely, no power consumption and no vibration. The energy *I* emitted from a surface plate of the radiator at temperature T with unit cross section is given by the Stefan–Boltzmann law, $I = \varepsilon \sigma T^4$, where σ is the Stefan–Boltzmann constant and ε is the emissivity of the surface material. Since the radiator receives thermal energy from other bodies in the field of view of the radiator, it is essential that there is no appreciable hot object in the field of view. In contrast, lunar explorer SELENE (Kaguya) (Hasebe *et al.*, 2008) and Mercury explorer MESSENGER (Goldsten *et al.*, 2007) employed Stirling cryocoolers, which need some power and cause some vibration but make the design of the cooling system easy — particularly for missions on an orbit near the hot planetary body.

Figure 3.1 shows the mass attenuation coefficient of germanium for gamma rays with the energy E_g from 0.001 to 20 MeV. The coefficient increases with each decrease in E_g below several MeV, except drops corresponding to the binding energy of a germanium atom. The drop near $E_g = 10$ keV is due to the binding energy of K-shell electrons of 0.0111 MeV; photons with less energy than this never ionize the K-shell electron. The drop near 0.001 Mev corresponds to the energies of the L_1, L_2, and L_3 shells of about 1.2–1.4 MeV.

Figure 3.2 shows a sample of pulse height spectra obtained by a small Ge gamma-ray detector with a liquid-nitrogen refrigerator for checking

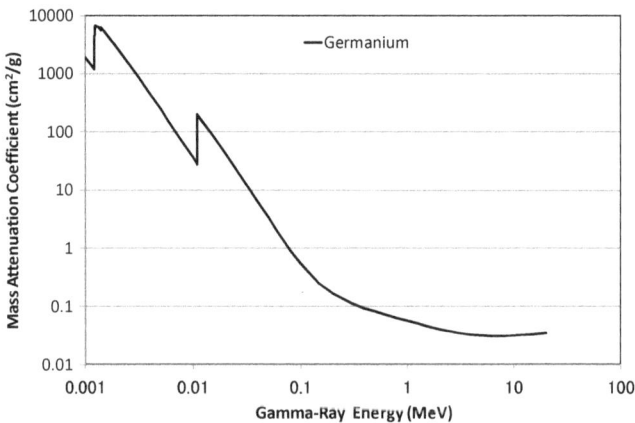

Figure 3.1: Mass attenuation coefficient for gamma rays in germanium. The data are presented in a mass attenuation table at NIST. (http://physics.nist.gov/PhysRefData/XrayMassCoef/tab3.html).

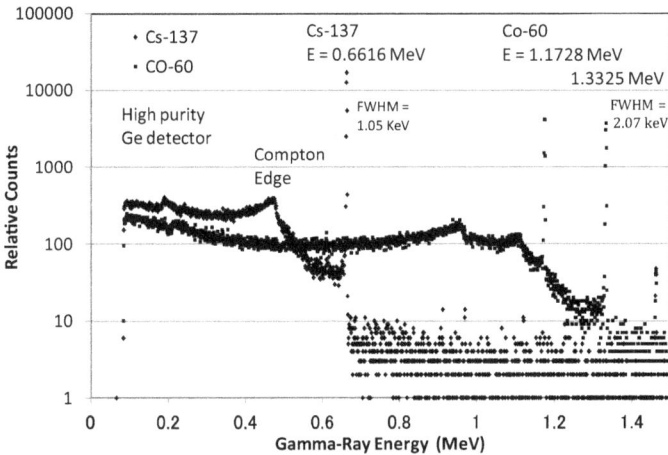

Figure 3.2:   Typical pulse height spectra obtained with a high purity germanium gamma-ray detector for sources of ^{137}Cs and ^{60}Co. The size of the germanium crystal is 49 mmϕ × 38 mm. The relative efficiency of the detector is 12%.

sources of ^{137}Cs and ^{60}Co. The peak for the ^{137}Cs source is due to a gamma ray with the energy of 0.6616 MeV. This gamma ray scatters electrons with the maximum energy of 0.478 MeV through Compton scattering. Thus, the energy of scattered electrons distributes below the edge, called the Compton edge, at 0.478 MeV, as clearly seen in the spectrum. The two peaks for the ^{60}Co source are due to gamma rays with energies of 1.333 and 1.173 MeV. The energy resolution is about 1.05 keV for 0.662 MeV gamma rays and 2.07 keV for 1.333 MeV gamma rays in full width at the half maximum (FWHM). These energy resolutions are in some degree worse than the best resolutions for the respective energies as shown in Table 3.1. The small peak at 1.46 MeV is due to natural radioisotope ^{40}K.

3.1.3 *Scintillation detectors*

In some transparent materials, scintillation photons are emitted from excited atoms by the ionizing radiation of charged particles or gamma rays. The number of scintillation photons provides information regarding the energy deposited in the material by the radiation. Materials with good properties for the detection of the scintillation photons produced by

Table 3.2: Some properties of inorganic scintillators.

Scintillator	Atomic number	Density (g/cm^3)	Max. wave length (nm)	Decay time (μs)	Photon yield (photons/ MeV)	Best resolution for 662 keV gamma ray
NaI(Tl)	11/53	3.67	415	0.23[b]	66,000[a]	7.1%[c]
CsI(Tl)	55/53	4.51	540	0.68–3.34[b]	67,600[a]	5.7%[c]
BGO	83/32/8	7.13	480	0.3[b]	11,900[a]	9.1%[c]
LaBr$_3$(Ce)	57/35	5.1	370	0.03[d]	80,200[a]	2.9%[d]
SrI$_2$(Eu)	38/53	4.59	435	1.2[e]	1,20,000[e]	3%[e]
CeBr$_3$	58/35	5.2	371[f]	0.017[f]	68,000[f]	3.6%[f]

[a] Sasaki *et al.*, 2010; [b] Knoll, 1999; [c] Sakai, 1987; [d] van Loef *et al.*, 2002; [e] van Loef *et al.*, 2009; [f] Shah *et al.*, 2005.

ionizing radiation are used as scintillation detectors or scintillators. These scintillation photons are converted into electrons with proper photo-detectrors. Table 3.2 summarizes the important properties of some inorganic scintillators for planetary gamma-ray spectroscopy.

The wavelength of the scintillation photon restricts the choice of photodetector. For example, we may, in general, choose a photomultiplier (PMT) for most scintillators with various wavelength spectra. The Si photodiode has sensitivity for photons with relatively long wavelengths and that are sometimes coupled with a CsI(Tl) scintillator. Recently, silicon PMT (Buzhan *et al.*, 2003) has been developed as a new photodetector.

The decay time is the time for the excited states to decay into the ground states in the scintillator. To obtain the signal for the total number of photons produced in the scintillator, an integration time that is longer than the decay time is needed. The decay time of the scintillator limits the counting rate, so a scintillator with a short decay time may be a better option for high-count-rate measurement.

The photon yield is the average number of photons produced by absorbing the total energy of a 1 MeV gamma ray, and is an index of the signal amplitude of the scintillator. Of course, a large photon yield is desirable for scintillators. The photon yield may be compared with the

ε value in a semiconductor. For example, a NaI(Tl) scintillator produces 66,000 photons for a 1 MeV gamma ray and, in most cases, the photons are converted to photoelectrons by a PMT with a quantum efficiency (Q_e) of 0.1–0.3, where the efficiency of photon collection (F_c) by the PMT is typically 0.5–0.7. Assuming that $Q_e = 0.2$ and $F_c = 0.6$, we obtain 7900 photoelectrons in a NaI(Tl) scintillator for a 1 MeV gamma ray. On the other hand, we obtain $10^6/\varepsilon = 3.36 \times 10^5$ electron–hole pairs for a 1 MeV gamma ray in Ge semiconductor detectors.

Energy resolution is an important factor of scintillators. LaBr$_3$, SrI$_2$(Eu), and CeBr$_3$ are recently developed scintillators, which have good energy resolutions. In this table, a LaBr$_3$(Ce) scintillator exhibits the best energy resolution of 2.9%. However, lanthanum naturally contains 0.09% ^{138}La, which is a radioactive isotope with a half-life of 1×10^{11} years. This isotope emits gamma rays with energies of 1436 or 789 keV. Camp *et al.* measured the gamma ray from LaBr$_3$ (Ce) scintillators and found that the activity of ^{138}La was 1.5 Bq/cm^3 (Camp *et al.*, 2016). The energy (1436 keV) is quite close to the energy of gamma rays from ^{40}K (1462 keV). This might be a serious problem, since ^{40}K is often an important element in planetary exploration. CeBr$_3$ scintillators with a short decay time or SrI$_2$(Eu) scintillators with a good energy resolution will be important for nuclear planetology in the future.

Figure 3.3 shows the mass-attenuation coefficient of a NaI(Tl) crystal for gamma rays as a function of the gamma-ray energy (E_g). The photoelectric effect is the most significant at $E_g < 0.2$ MeV, while pair creation is the most significant at 6 MeV $< E_g$ and Compton scattering is the most significant at 0.2 MeV $< E_g < 6$ MeV. The mass-attenuation coefficient of NaI(Tl) is dominated by iodine since the atomic number of iodine is larger than that of sodium. The steep step near $E_g = 0.033$ MeV is due to the K shell of iodine.

Figure 3.4 shows a sample of pulse-height spectra obtained by the NaI(Tl) scintillator for checking sources of ^{137}Cs and ^{60}Co. The energy resolution in FWHM is about 8% for a 661.6 MeV peak from ^{137}Cs, which is fairly close to the best value in Table 3.2. The energy resolution in scintillators is worse than that in germanium semiconductors mainly due to the following two factors (Moszynski *et al.*, 2002; Prescott *et al.*, 1969):

(1) Photon yields: the number of photons produced by the gamma ray per 1 MeV in the scintillators are smaller than the number of ion–hole pairs

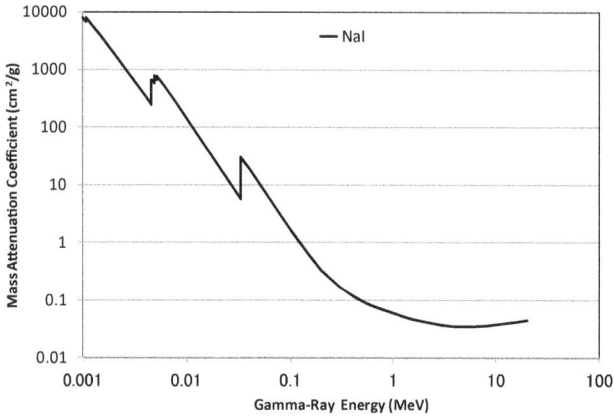

Figure 3.3:   Mass-attenuation coefficient for gamma rays in a NaI(Tl) crystal. This is drawn from the mass-attenuation coefficients of Na and I presented in a mass-attenuation table at NIST (http://physics.nist.gov/PhysRefData/XrayMassCoef/tab3.html).

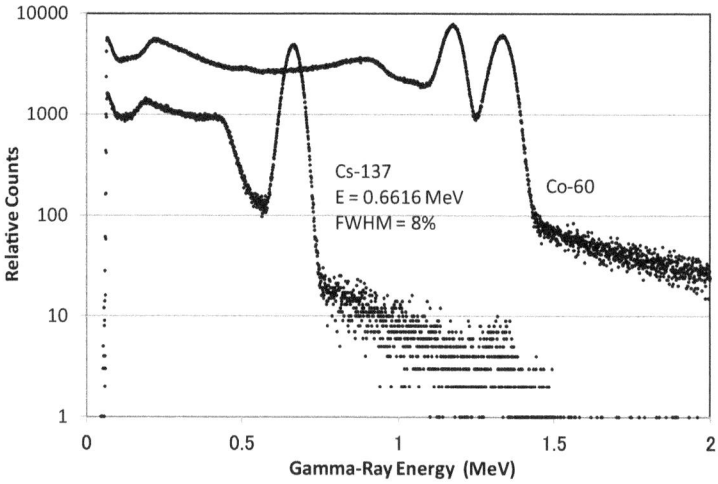

Figure 3.4:   Sample of pulse-height spectra obtained with the NaI(Tl) scintillator for ^{137}Cs and ^{60}Co gamma-ray sources.

in the germanium semiconductor for gamma rays with the same energy. As shown in Table 3.2, the photon numbers in these scintillators range from 11,000 to 120,000 photons/MeV. The number of electron–hole pairs is 336,000 for a 1 MeV gamma ray in the germanium semiconductor. So, the statistical fluctuation in photon numbers is relatively larger than that in the number of ion–hole pairs. Furthermore, we need to convert these photons into electrons through some photodetectors including PMTs, photodiodes, or silicon photomultipliers. The conversion efficiency of photons to electrons is less than unity. This introduces additional statistical fluctuation in relative energy resolution.

(2) Electron response: the efficiency of photon production by an energetic electron may be dependent on the energy of the electron, which is called the "electron response". The peak signal is due to the sum of electrons produced in the scintillator in an event where the total energy of an initial gamma ray is absorbed in the scintillator. Although the sum energy of the electrons equals the gamma-ray energy, the respective combination of electron energies is different from event to event. So, the electron response causes the fluctuation in the number of scintillation photons produced in the scintillator. In most scintillators, more or less, such an electron response is reported.

3.2 Neutron Detector

Neutrons have no electric charge, so they are detected when they interact with nuclei in the detector material and produce charged particles. Slow neutrons in general have a large cross section of interaction with nuclei. The kinetic energy recoiled by the slow neutron is small. So, slow neutrons are detected through the detection of charged particles or gamma rays emitted through a neutron-induced reaction with positive Q value, where the Q value is the mass difference in energy scale between the total mass of interacting particles before and after the interaction. On the other hand, fast neutrons may scatter nuclei elastically with energy ranging from zero to the energy of the neutron, or may scatter inelastically. The fast neutrons may be detected by measuring the energy of recoil particles or gamma rays emitted from the excited atom.

3.2.1 *Thermal neutron detection*

Thermal neutrons are mainly detected through neutron absorption by nuclei followed by the emission of charged particles where energy

(Q value) is released. The cross section of neutron absorption is, in general, a decreasing function of neutron energy — except at the resonance energies, where the cross section is appreciably large. The three isotopes, ^{10}B, ^6Li, and ^3He, have large cross sections of neutron reaction and large Q values. These are important for thermal neutron detection.

18.8% of naturally occurring boron is made up of the ^{10}B isotope. In interacting with neutrons, ^{10}B produces ^7Li and an alpha particle:

(i) $^{10}B + n \rightarrow\ ^7Li + \text{alpha} + 2.792$ MeV, or
(ii) $^{10}B + n \rightarrow\ ^7Li* + \text{alpha} + 2.310$ MeV,
       $^7Li* \rightarrow\ ^7Li + \gamma\,(0.48\ \text{MeV})$,

where ^7Li and $^7Li*$ are the ground state and excited state of ^7Li, respectively. This interaction has a larger Q value of 2.78 MeV, which is divided between the alpha particle and the ^7Li nucleus. The energy of the alpha particle and the ^7Li nucleus are measured using an ionization chamber, a proportional counter, or a scintillation detector. The ratio of the energy given to the alpha particle and the ^7Li nucleus is inversely proportional to the respective mass in the case of the absorption of a thermal neutron, where the kinetic energy of the neutron is negligible. In the case of (i), the energy of the alpha particle and that of the ^7Li is 1.777 MeV and 1.015 MeV, respectively. Figure 3.5 shows the cross section of ^{10}B for a neutron reaction as a function of neutron energy.

One neutron detector is a BF_3 proportional counter with ^{10}B enriched to about 90%. BF_3 is a gas detection medium and is capable of gas multiplication at a high electric field. The energetic Li and alpha particle ionize BF_3 gas and the ionized electrons are detected. Boron-doped plastic scintillators are also available. Boron-doped plastic scintillators were employed in Lunar Prospector to detect neutrons (Feldman *et al.*, 1999).

The ^6Li isotope makes up 7.5% of natural lithium. Interacting with neutrons, ^6Li produces ^3H and an alpha particle:

$$^6Li + n \rightarrow\ ^3H + \text{alpha} + 4.79\ \text{MeV}.$$

The energy of ^3H and the alpha particle are measured by a scintillator or a semiconductor detector. Figure 3.6 shows the cross section of ^6Li for neutron reaction as a function of neutron energy. The Dawn mission carries

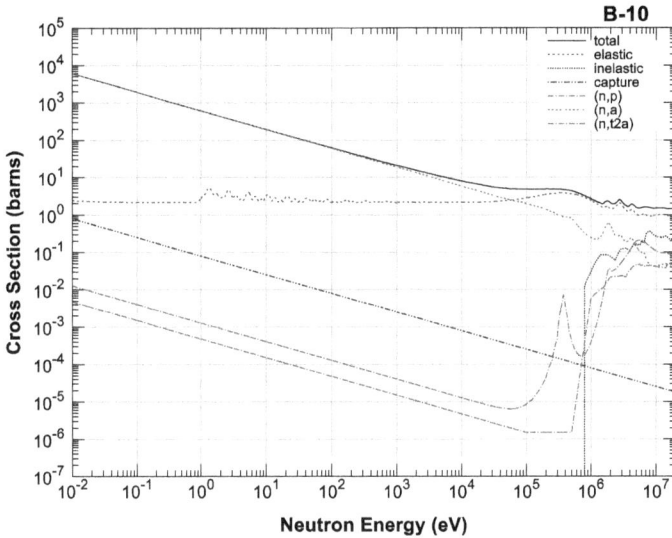

Figure 3.5:   Cross section of ^{10}B for neutron reaction (Shibata *et al.*, 2011). The dominant reaction is (n, α) reaction for neutron energy E_n less than 10^4 eV and is elastic scattering for E_n larger than 10^5 eV.

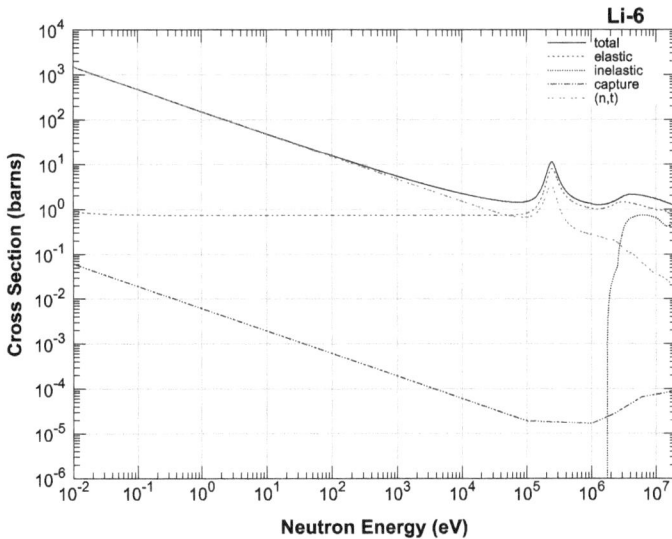

Figure 3.6:   Cross section of ^{6}Li for neutron reaction (Shibata *et al.*, 2011). The dominant reaction is a (n, α) reaction for neutron energy (E_n) less than 10^4 eV and is elastic scattering for E_n larger than 10^5 eV.

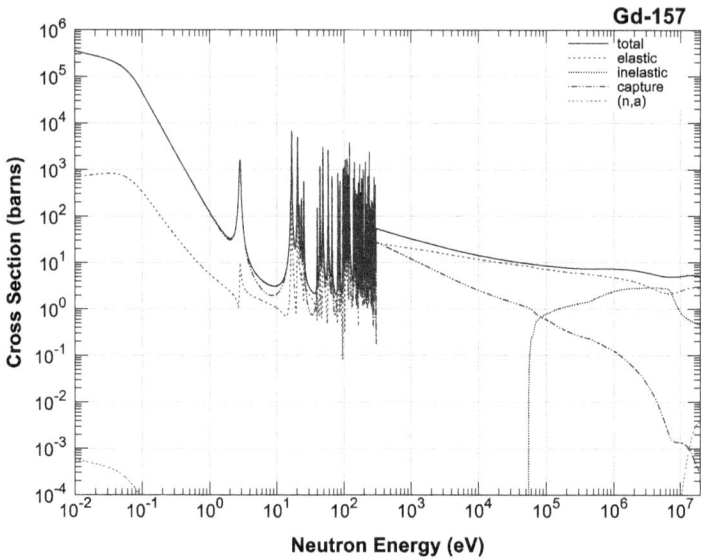

Figure 3.7:   Cross section of ^{157}Gd for neutron reaction (Shibata *et al.*, 2011). The dominant reaction is the capture reaction for neutron energy E_n less than 10^1 eV, and is elastic scattering for E_n larger than 10^3 eV.

^{6}Li-loaded glass scintillators as one of the neutron detectors (Prettyman *et al.*, 2003).

The ^{157}Gd isotope makes up 15.7% of natural gadolinium. After capturing a neutron, ^{157}Gd produces a stable ^{158}Gd isotope. Figure 3.7 shows the cross section of ^{157}Gd for neutron reaction as a function of neutron energy. The capture cross section at thermal neutron energy is larger than most other atoms.

As the cross section of ^{157}Gd is as large as 255 kb at thermal energy, 70% of the thermal neutron is stopped with a 10-μm thick foil of natural gadolinium. A gadolinium sheet is employed in the neutron spectrometer onboard the Lunar Prospector to reject thermal neutrons for a particular detector.

3.2.2 *Fast neutron detection*

Fast neutrons are mainly detected by detecting recoil nuclei. The neutron transfers the energy to ^{1}H most efficiently through elastic collision since ^{1}H almost equals one neutron in mass. Figure 3.8 shows the cross section

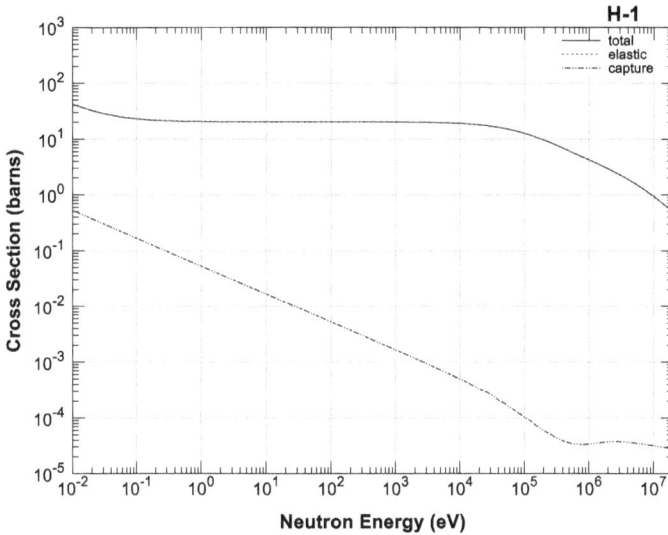

Figure 3.8:   Cross section of ^{1}H for neutron reaction (Shibata *et al.*, 2011). The dominant reaction is elastic scattering for neutrons with energy from thermal to 10 MeV.

of ^{1}H for a neutron reaction. For thermal neutrons, the capture cross section of ^{1}H is not negligible.

3.3 Active Fluorescence X-ray Spectrometer

X-rays are emitted by atoms with electron transition from an orbit to the orbit of lower energy. The principle of photon emission is the same as that in flame reaction. Usually these photons with energy over 0.1 keV are called X-rays. Since the energy of X-ray is specific to the emitting atom, these X-rays are called characteristic X-rays. Table 2.3 shows the energy of Kα X-rays emitted by atoms from ^{4}Be to ^{36}Kr. The energy of a Ka X-ray increases with the atomic number. The differences in the X-ray energy between adjacent atoms are about 200 eV around $Z = 4$ and about 700 eV around $Z = 36$. To identify the energy of a Kα X-ray, the energy resolution of the X-ray spectrometer should be enough better than 200 eV or 700 eV, depending on the target atoms it needs to to identify. Energetic X-rays and charged particles can excite atoms into the higher-energy levels needed to emit X-rays. In planetary missions using the rover so far, radioactive sources have been used.

3.3.1 *Artificial X-ray generation to induce fluorescence X-ray*

X-rays emitted from atoms excited by X-rays are called fluorescence X-rays and are used in the identification of the atoms. Any atom is efficiently excited by a photon with energy slightly higher than the excitation energy, as the steep step of mass-attenuation coefficient at nearly 0.011 MeV shows the energy of a K X-ray from a germanium atom (as sown in in Fig. 3.1). To artificially induce fluorescence X-rays, intense X-rays are necessary, and these are possible in landing or rover missions.

Recently, studies of pyroelectric crystals for X-ray generation have been conducted. The crystals induce an electric charge when the temperature changes. Free electrons are accelerated by the electric field by the induced charge and hit some target or the crystal, depending on the phase of cooling or heating, to excite the target atoms. Kusano *et al.* (2014, 2016) generated about 10^7 X-ray photons per second with a pyroelectric crystal.

3.3.2 *X-ray detectors*

In X-ray spectroscopy, various detectors are applied, such as proportional counters, Si PIN diodes, and silicon drift detectors (SDD). Proportional counters (Adler *et al.*, 1972a) were employed on the Apollo 15 and 16 missions. The ratio of atoms, Mg/Si and Al/Si, were obtained over the area near the equator of the Moon. In recent planetary explorations, Si PIN diodes and Si SDDs are often employed because of their better energy resolution. The detector with a large detection area gives a high detection efficiency, but at a cost (in most cases) of a lower energy resolution due to its large electrostatic capacitance. The capacitance can be minimized in an SDD. Recently, an SDD with a detection area of 25 mm^2 and energy resolution better than 125 eV for 5.9 keV X-ray was made commercially available by Amptec Inc.

Chapter 4

Numerical Simulation
and Laboratory Experiment

The Monte Carlo N-Particle eXtended (MCNPX) code and the Los Alamos High-Energy Transport (LAHET) code system (LCS) have been widely used for planetary applications associated with radiation monitoring and particle production (McKinney *et al.*, 2006). Masarik and Reedy (1996) studied a few numerical simulations using the LCS to simulate the interaction of cosmic ray particles with the atmosphere and surface of a planetary system. Using the MCNPX code Kim *et al.* (2007; 2010) conducted similar studies and investigated production of gamma rays and neutrons on planetary surfaces and extraterrestrial materials. For the past few decades, gamma rays and neutrons from planetary surfaces have been investigated using remote-sensing techniques. These radiations are mainly produced by nuclear reactions between incoming cosmic rays and planetary materials. It is important to estimate the production rates of gamma rays or neutrons emitted from a targeted planetary surface prior to a planetary investigation. The MCNPX code has been widely used to estimate the production rates of these radiations and cosmogenic nuclides from planetary surfaces and extraterrestrial materials in space. This chapter introduces the production of radiations and radionuclides in a laboratory environment, the numerical simulations used for these experimental cases, and the theoretical values of gamma-ray production on planetary surfaces.

4.1 Production of Radionuclides in Extraterrestrial Materials and at Lunar Surface

4.1.1 *Production rates of cosmogenic nuclides*

In the early 1960s, the moon probe Ranger 3 was used to measure gamma rays in space (Arnold *et al.*, 1962) and to investigate the abundance of radioactive and stable nuclides produced by cosmic rays in meteorites, where the half-lives of these nuclides ranged from 16 days to 1.2×10^9 years (Arnold *et al.*, 1961). The production of measureable quantities of He and other isotopes by cosmic-ray bombardment in meteorites was first predicted by Bauer (1948) and his observation was reported by Paneth *et al.* (1952). Arnold *et al.* (1961) investigated the production rates of cosmogenic nuclides and recorded the intensity of cosmic rays in mete- orites (Figs. 4.1 and 4.2). Radioactive nuclides in meteorites are compared

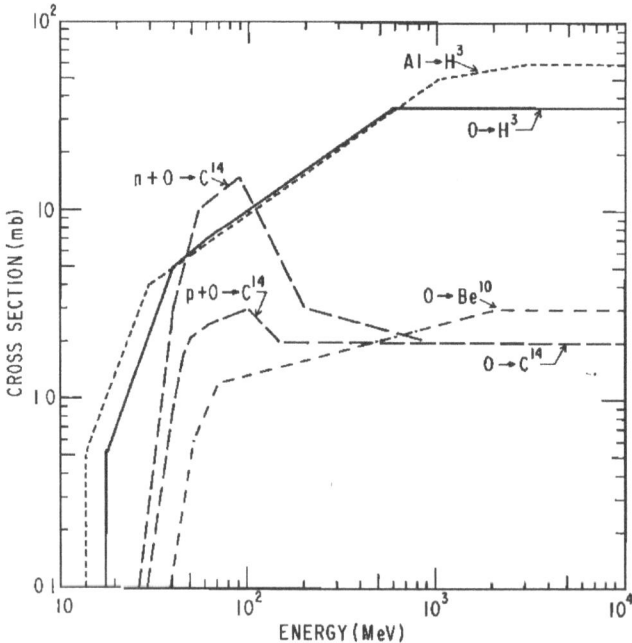

Figure 4.1:   The excitation functions of the production of ^{14}C, ^{10}Be, and ^{3}H from oxygen and ^{3}H from aluminum by the interaction of GCR and SCR particles (Reedy and Arnold, 1972).

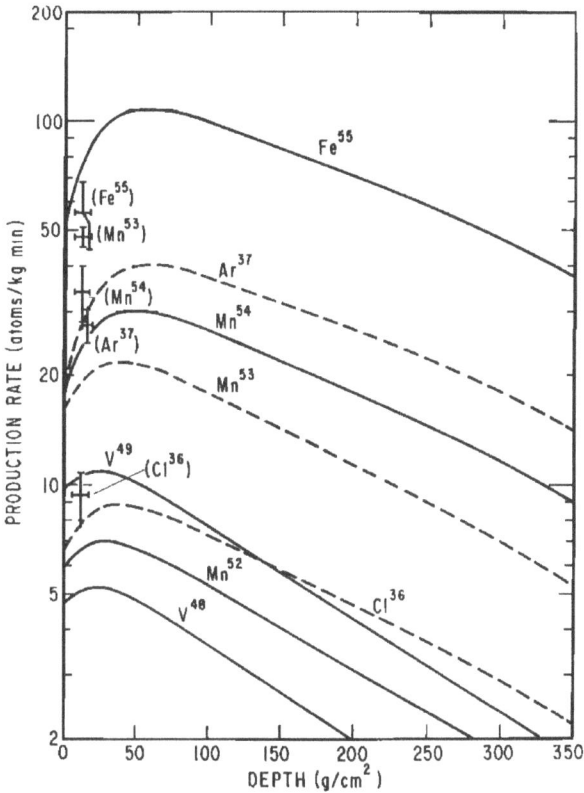

Figure 4.2:   The calculated production rates of GCR particles as a function of depth.

with production cross sections in bombardment of Fe with 730 MeV protons. Arnold *et al.* (1961) also reported various cross sections assumed for the normalization of excitation functions. The fundamental works in both experimental and theoretical fields of cosmochemistry by James R. Arnold, Devandra Lal, and M. Honda in the early years between the 1950s and 1960s brought on a new era in cosmochemistry research and eventually opened a new field: nuclear planetology. Further investigation on excitation functions of cosmogenic nuclides and their production as a function of depth from the Moon for both solar cosmic ray (SCR) and galactic cosmic ray (GCR) particles was accomplished by Reedy and Arnold (1972).

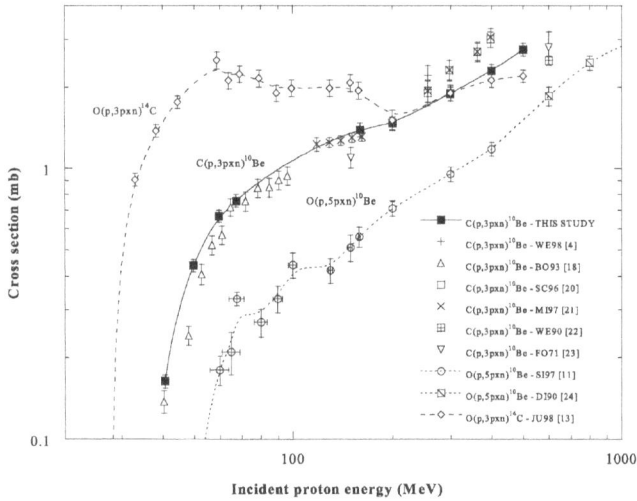

Figure 4.3:   The measured cross sections for ^{10}Be production from both carbon and oxygen and for ^{14}C production from oxygen (Kim *et al.*, 2002).

Proton cross sections of long-lived nuclides, ^{10}Be and ^{26}Al, and short-lived nuclides, ^{7}Be and ^{22}Na, from elements found in lunar rocks were reported for SiO_2 targets as a function of proton energies (Sisterson *et al.*, 1997). Proton cross sections of ^{10}Be for natural carbon targets were reported for proton energies ranging from 40.6 to 500 MeV (Kim *et al.*, 2002). The general trend of a proton cross section as a function of proton incident energy demonstrates that the proton cross sections of ^{10}Be for carbon targets are higher than those for oxygen targets and are lower than the cross sections of ^{14}C for oxygen targets at the same incident proton energy (Fig. 4.3) (Kim *et al.*, 2002). Using 5- to 25-cm-thick spheres of diorite, gabbro, and iron targets, nearly 100 target elements were investigated. The secondary neutrons were estimated, and the depth dependence of spallation reactions in a spherical thick diorite target irradiated by protons was investigated (Fig. 4.4) (Leya and Michel, 2011).

These cosmogenic nuclides need to be understood for various geological applications related to their production in space and terrestrial environments. Accurate estimations of cosmogenic nuclide productions induced by both protons and neutrons are important in the terrestrial application of geomorphological studies as well as climate changes

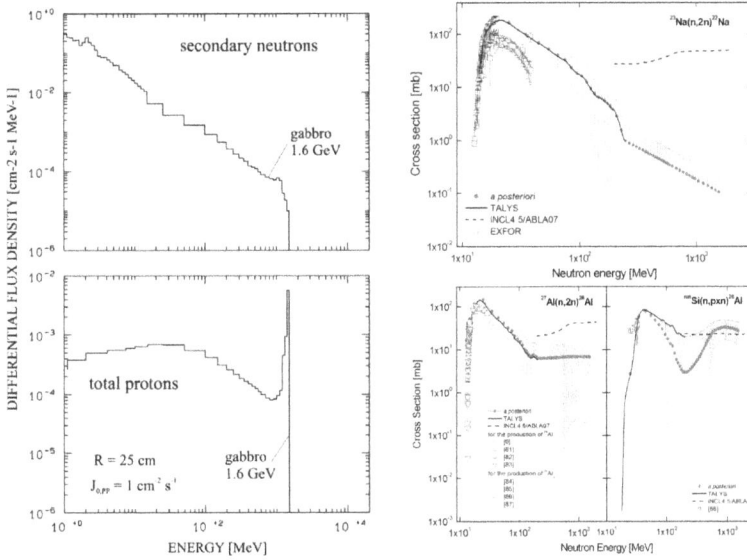

Figure 4.4:   Differential flux densities for neutrons and protons in the center of 25-cm-thick gabbro sphere irradiated with 1.6 GeV protons (left figures) and neutron-induced ^{22}Na and ^{26}Al cross sections as a function of the neutron energy (right figures) (Leya and Michel, 2011).

through time (Kim *et al.*, 2007; Yoshimori *et al.*, 2003). Also, irradiation history and the terrestrial age of meteorites can be obtained by examining cosmogenic nuclide abundance. Lunar samples from the Apollo missions were also investigated for these cosmogenic nuclides, and the results of their abundance help us understand both GCR and SCR particle fluxes in the lunar environment (Fink *et al.*, 1998; Kim *et al.*, 2010). The MCNPX numerical simulations of production rates for ^{10}Be, ^{26}Al, and ^{14}C in extraterrestrial matter such as Kynahinya and on the lunar surface confirm that the average effective proton fluxes were close to the published value of 4.8 protons/sec/cm^2 given by Reedy *et al.* (1993) (Fig. 4.5), Kim *et al.* (2010), and Reedy and Masarik (1994).

4.1.2 *Production rates of gamma rays and neutrons*

The production of gamma rays has been investigated following inelastic neutron scattering or neutron capture interactions (Yates *et al.*, 1978;

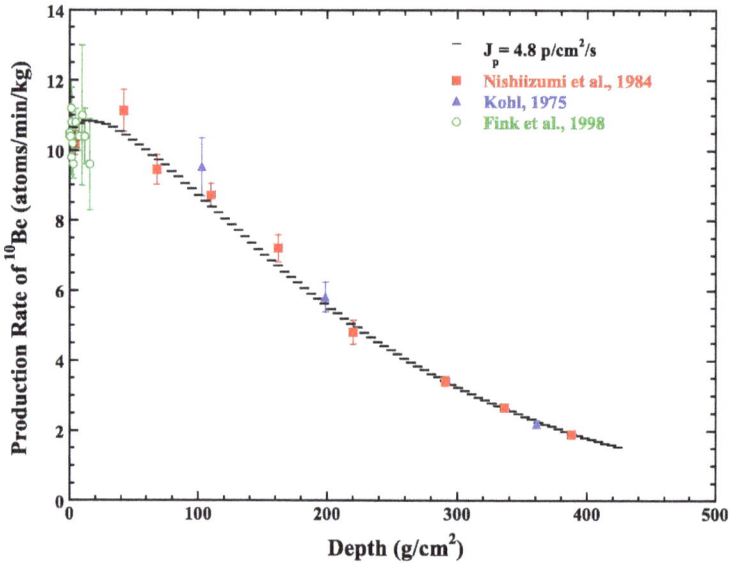

Figure 4.5:   Calculated production rates of ^{10}Be as a function of depth for the Apollo 15 drill core and the surface of Apollo 17 site (Kim *et al.*, 2010).

Zwittlinger, 1973). In neutron scattering, a 14-MeV neutron was used for elemental analysis by Schrader and Stinner (1961). Yates *et al.* (1978) used 1.5-, 2.5-, and 3.5-MeV neutrons to bombard an iron target. Many more gamma-ray lines are produced for 3.5-MeV neutron energies. Numerous sets of experimental data are available on this subject. On both planetary surfaces and extraterrestrial materials, iron is in abundance. Gamma-ray activation analysis was performed to confirm the presence of iron. Neutron-capture gamma-ray activation analysis of refined steel provided main gamma-ray peaks from the elements Fe, Cr, Ni, Mn, W, and Co. The energies of these gamma rays vary from 5.311 to 8.996 MeV (Zwittlinger, 1973) and are useful in the identification of gamma-ray lines for planetary remote-sensing applications (Table 4.1). Since the Apollo lunar program, planetary gamma-ray spectroscopy has been adapted for elemental analysis on the Moon, Mars, Mercury, and asteroid bodies (Banerjee and Gasnault, 2008; Boynton *et al.*, 2002; 2004; Brückner and Masarik, 1997; Evans *et al.*, 2001; 2012; Lawrence *et al.*, 2000; Peplowski *et al.*, 2011; 2012a; 2012b; 2013; 2015; Reedy and Masarik, 1994; Rhodes *et al.*, 2011; Trombka *et al.*, 2001; Usui *et al.*, 2010). Early investigation

Table 4.1:   Main peaks for the measured elements Fe, Cr, Ni, Mn, W, and Co.

	Useful peaks for analysis		Peaks with interferences		
Element	Energy (MeV)	Intensity	Energy (MeV)	Intensity	Element
Fe	5.920	8.29	7.277	4.6	Mn
	6.018	8.08	7.628	27.19	Cu
			7.643	22.14	Cu
	5.260	1.13	6.872	1.03	Co, V
	5.610	3.46	7.364	6.73	Pb
	6.000	2.27	7.929	11.41	Cu
Cr	6.120	1.66			
	6.644	5.29			
	7.097	3.88			
	8.499	5.50			
	8.881	24.14			
	5.311	1.11	5.814	2.34	Cd
	6.103	2.08			
	6.579	1.95			
	6.835	11.91			
Ni	7.537	4.94			
	7.814	9.04			
	8.114	3.47			
	8.525	18.74			
	8.996	18.74			
	6.107	1.90	5.530	6.94	V
Mn	6.788	3.46	7.164	6.06	V
	7.061	11.35	7.274	3.08	Fe
	7.248	12.05			
W	5.250	4.13			
	6.194	4.39			
	5.270	1.11	5.742	1.98	V
	5.603	1.07	5.926	1.73	Fe
	5.660	6.21	6.877	7.77	V
	5.976	6.49	7.056	1.65	Mn
Co	6.486	6.29			
	6.706	7.22			
	6.985	2.82			
	7.214	4.56			
	7.491	2.91			

Source: Zwittlinger, (1973).

of gamma rays on planetary surfaces was accomplished by Reedy and Arnold (1972, 1973) by studying expected gamma-ray emission spectra from the lunar surface as a function of chemical composition. They listed gamma-ray emission rates from both natural radioactive and major elements from the Moon under the conditions of the Apollo 15 and 16 experiments. The expected fluxes of gamma rays were calculated with four lunar compositions and one chondritic chemistry, taking into consideration the important emission mechanisms of natural radioactivity, inelastic scatter, neutron capture, and induced radioactivity. These emission rates of gamma rays have been widely used for various gamma-ray spectrometer (GRS)-associated planetary missions. Importantly, Reedy and Frankle (2002) compiled all prompt gamma-ray data from the radioactive capture of thermal neutrons by elements from hydrogen to zinc. This publication provides an essential literature data set compiled up to August 2000 for all major gamma-ray lines produced by neutron capture interactions in planetary gamma-ray remote-sensing applications (Reedy and Frankle, 2002). They used the published thermal capture cross sections to convert the isotopic gamma-ray intensities to elemental abundances. The evaluated energies and absolute intensities of these capture gamma rays are presented by elements concerned.

4.2 Numerical Simulation and Thick-Target Experiment

For the simulation of the interaction of the GCR particles with matter, a few thick-target experiments have been conducted with respect to gamma-ray production on planetary surfaces (Paul *et al.*, 1995). In regard to 2.1-GeV protons and 800-MeV/nucleon alpha particles, the spatial distribution of low-energy secondary neutrons successfully determined neutron capture reaction products on the irradiated targets with a size of $100 \times 100 \times 200$ cm and a mass of about 16,000 kg. Relative activities of ^{56}Mn, ^{198}Au, and ^{187}W as a function of depth were investigated (Englert *et al.*, 1987). The results of this study are applied in evaluating the subsurface low-energy neutron fluxes in extraterrestrial objects. Comparisons with similar results from other thick-target experiments demonstrate the advantages of the thick-target arrangement used (Englert *et al.*, 1987). Another thick-target experiment was conducted prior to the Mars Observer mission (Brückner *et al.*, 2011). To investigate the production of gamma rays, a thick-target experiment was performed from 1991 to 1993 using two proton energy sources, simulated Martian soil and an iron target. This planetary gamma-ray simulation experiment was carried out at the French accelerator, Laboratoire National SATURNE (CEACNRS), which was based at the Centre d'études nucleaire, Saclay, France. This experiment revealed possible gamma-ray production in Martian soil. The comparison of gamma-ray spectra of a basalt target and a water-rich target including volatiles was confirmed to be a very important step toward planetary gamma-ray spectroscopy on Mars as well as understanding lunar gamma rays. Gamma-ray production on the simulated Martian soil was investigated using the MCNPX code by Masarik and Reedy (1996) prior to the Mars Odyssey mission. For both numerical simulation and experimental study, GCR energy was considered. As an input parameter, the GCR flux with differential energy from 0 to 30 GeV was used, whereas for the thick-target experiment, both 1.5- and 2.5-GeV monoenergetic proton beams were used to irradiate the thick targets. The thick-target experiment confirmed that the gamma-ray spectra from both 1.5- and 2.5-GeV proton beams have produced similar numbers of gamma rays but with slightly different intensities. Figure 2.3 in Chapter 2 shows the GCR flux (Yamashita *et al.*, 2008). Maximum GCR proton lies at 800 MeV.

The experimental setting at CEACNRS, Saclay, for the proton irradiations is shown in Fig. 4.6. Table 4.2 shows the compositions of each target (Fabian *et al.*, 1996). A High-purity Germanium (HPGe) detector was used to measure the gamma rays produced for the four targets. The gamma-ray productions were reported in photon/cm^2 in this study. Figures 4.7–4.9 show the gamma-ray spectra for the basalt target 1. The thick-target gamma-ray spectra were important for predicting gamma-ray production on Martian surface (Figs. 4.7 and 4.8). Figure 4.9 shows the comparison of the thick-target spectrum with the Mars Odyssey GRS spectrum. This figure shows almost all Martian gamma rays could be identified with gamma rays obtained from the thick-target experiment (Brückner *et al.*, 2011).

Yamashita *et al.* (2006) performed bombardment experiments with thick targets made of Al and Fe respectively, using 230 MeV/nucleon helium and 210 MeV proton beams. The energy spectra of gamma rays

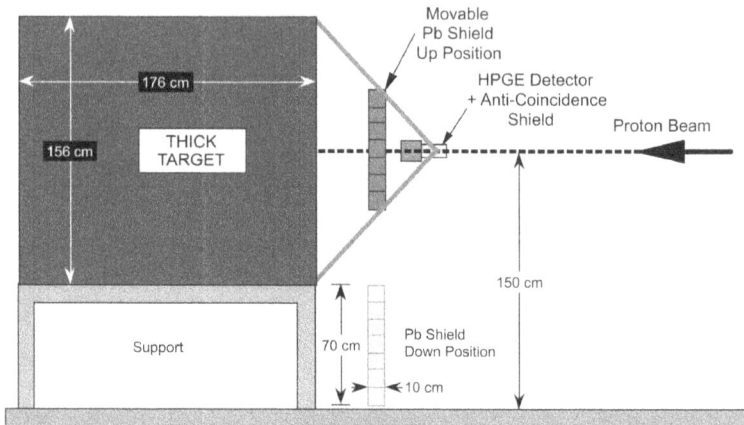

Figure 4.6:   Side view of the thick-target experiments. The movable lead is used to attenuate the target signal or measure the signals from the target and background (Brückner *et al.*, 2011).

Table 4.2:   Compositions of the thick targets.

Elements	Pure basalt [wt.%]	Abundance B (basalt + steel) [wt.%]	Abundance V [wt.%]	Abundance W [wt.%]	Abundance F [wt.%]
Si	20.08	18.32	20.0	19.6	—
Ti	1.60	1.46	1.6	1.6	—
Al	6.71	6.11	6.7	6.5	—
Fe	8.57	16.60	8.5	8.3	~99
Mn	0.14	0.22	0.1	0.1	~0.5
Mg	6.38	5.81	6.3	6.2	—
Ca	7.42	6.76	7.4	7.2	—
Na	1.86	1.69	1.8	1.8	—
K	1.23	1.12	1.2	1.2	—
P	0.21	0.19	0.3	0.3	—
Cl	—	—	0.74	0.76	—
S	—	—	2.4	2.4	—
C	—	—	0.65	2.3	—
H	0.42	0.38	0.45	0.73	—
O	45.40	41.35	41.8	40.9	—

Source: Brückner *et al.*, (2011).

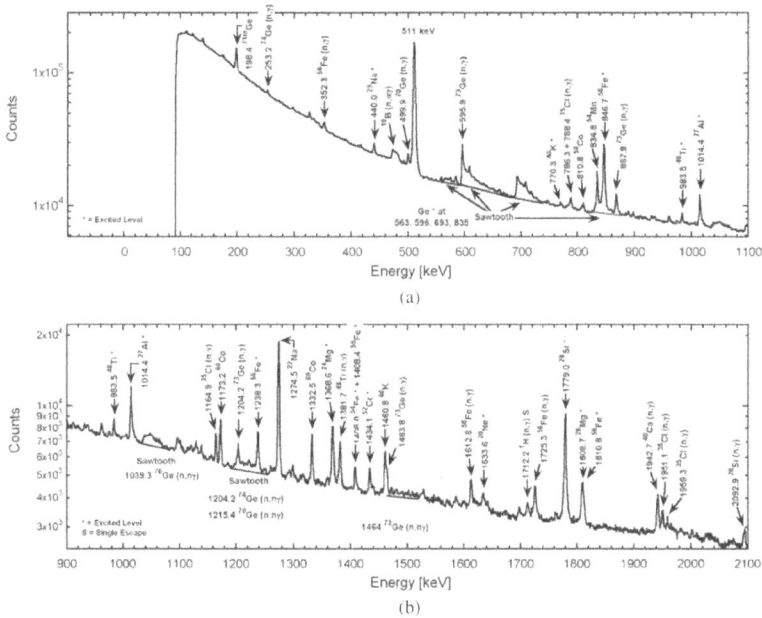

Figure 4.7: Gamma-ray spectra of basalt B1, irradiated by 1.5-GeV protons, target plus background, energy range (a) from 0 to 1100 keV and (b) from 900 to 2100 keV (Brückner *et al.*, 2011).

emitted from the iron thick target bombarded with the 230 MeV/nucleon He-beam are shown in Fig. 4.10. The energies and intensities of prompt gamma rays emitted by the inelastic scattering of energetic neutrons were measured by a HPGe detector. They presented the first experimental results concerning the emission of prompt gamma rays from thick targets irradiated by He, a component in galactic cosmic rays, since no foregoing work related to this has been done. We should note the role of He in planetary gamma-ray spectrometry; Helium has the capability of prompt gamma ray line emission through the inelastic nuclear reaction of its neutrons with the target material by a factor of 3.5 (on average), in comparison to its protons.

A numerical simulation of gamma-ray production on the thick target was performed using the Monte Carlo transport code by Fabian *et al.* (1996). The calculations were accomplished based on the Los Alamos LCS using the exact experimental setup. The ratios of gamma rays

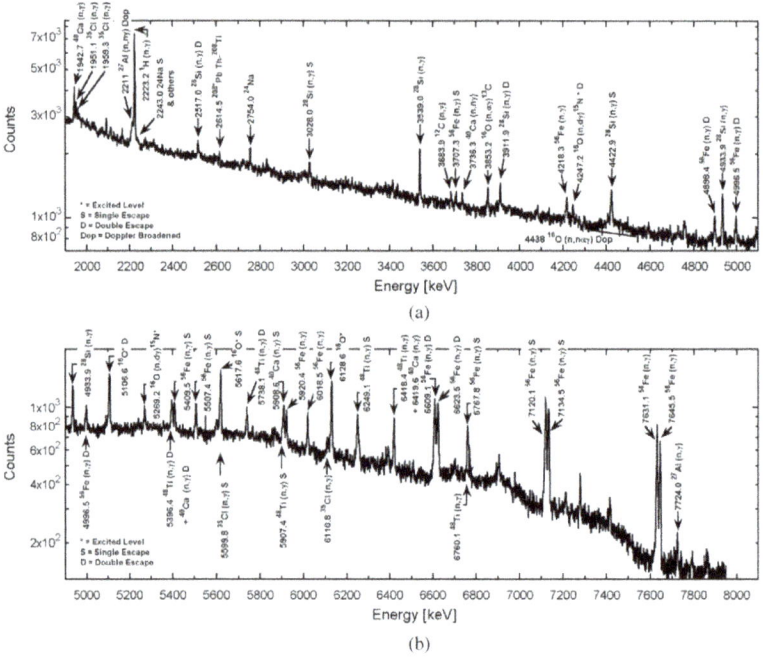

Figure 4.8:   Gamma-ray spectrum of basalt B1 (1.5 GeV), target plus background, energy range (a) from 1900 to 5100 keV and (b) from 4900 to 8100 keV (Brückner *et al.*, 2011).

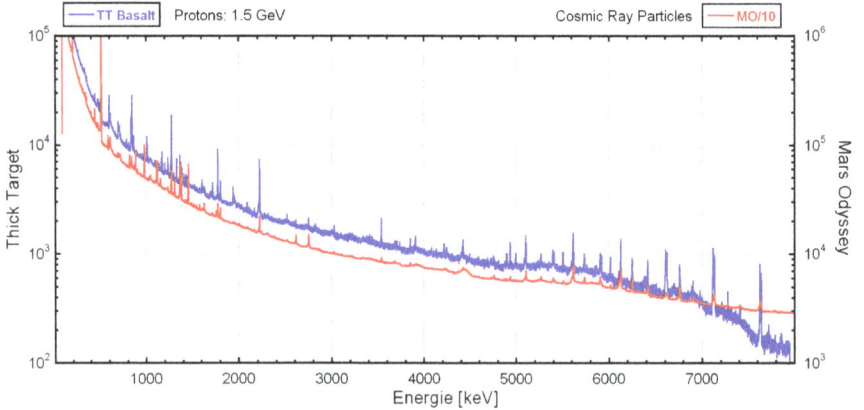

Figure 4.9:   Gamma-ray spectra from the basalt thick-target containing steel, irradiated with 1.5-GeV protons, and from the orbiting "2001 Mars Odyssey" gamma-ray spectrometer, accumulated over the entire Martian surface from 8 June 2002 to 13 June 2003. The Martian spectrum was scaled down by a factor of ten for ease of viewing. (Brückner *et al.*, 2011).

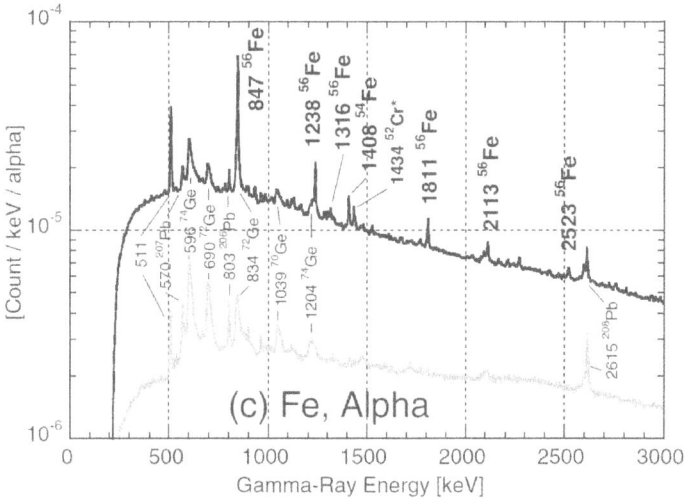

Figure 4.10:   Energy spectra of gamma rays emitted from iron thick target bombarded with a 230 MeV/nucleon helium beam. (Yamashita *et al.*, 2006).

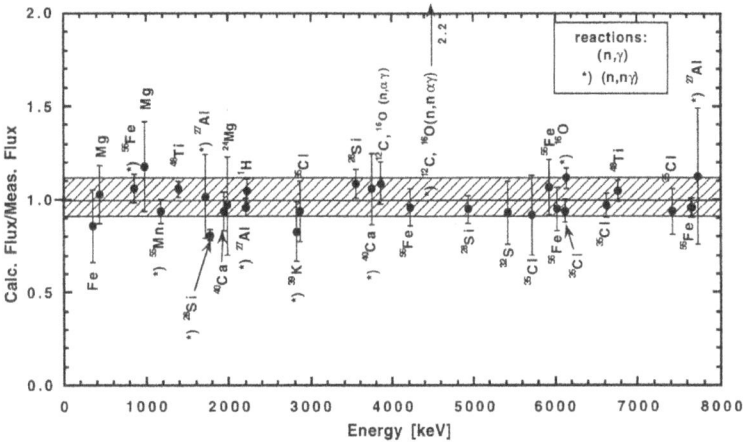

Figure 4.11:   Ratios of calculated-to-measure gamma-ray fluxes for selected, prominent lines from thick basalt target with the addition of chlorine, hydrogen, carbon, and sulfur (Fabian *et al.*, 1996).

produced from the exact experiment and numerical simulation were found to be mostly within 10% variations (Fig. 4.11), showing the reliability of the numerical calculation method using this Monte Carlo gamma-ray transport code.

Chapter 5

Nuclear Planetary Exploration

5.1 Apollo Gamma-Ray and X-Ray Spectrometer

5.1.1 *Apollo gamma-ray spectrometer*

Apollo gamma-ray spectrometers (AGRSs) were carried on Apollo 15/16 spacecraft, which were launched in 1971 and 1972, respectively (Harrington *et al.*, 1974). These missions covered about 22% of the surface of the Moon, all within 30° of the lunar equator. The AGRSs were 511-channel gamma-ray spectrometers, designed to determine the surface composition of the upper 30 cm of the Moon by detecting the 0.2–9 MeV line gamma rays. The main detector of an AGRS, as shown in Fig. 5.1 (a), consists of a 7.0 cm × 7.0 cm cylindrical crystal of NaI(Tl) with a 7.6-cm photomultiplier tube (PMT; RCA C31009) and a plastic mantle for anticoincidence rejection of charged particles. This crystal was hermetically sealed in a stainless-steel can of 7.6 cm in diameter and 8.6 cm in length. MgO powder was used as a reflector. The NaI crystal was covered by a 5-mm-thick borosilicate-glass optical window. Silicone grease was optically coupled with the crystal assembly to the PMT. An 8-mm-thick shield of Pilot-B plastic scintillator covered the NaI(Tl) on all sides, except for the crystal assembly, to the PMT interface to eliminate galactic cosmic ray signals. Events in the plastic scintillator produced a signal that inhibited the acceptance of a coincident pulse from the NaI(Tl). A 3.8-cm PMT (RCA C70114M) was coupled to the plastic shield. The AGRSs were mounted on the end of an 8-m boom. The detector weighed 3.3 kg and the electronics weighed 3.2 kg.

Although the AGRSs used low-resolution NaI(Tl) crystals, they could still produce maps of K, Ti, Fe, and Th on the Moon after a careful and complex deconvolution analysis of the spectra data with few clear peaks (Metzger, 1993; Metzger *et al.*, 1973). They showed that the anorthositic fragments seen in the soil sample collected by Apollo 11 are the dominant material of the lunar crust and that the highest concentrations of potassium (K), rare-earth elements (REE) and phosphor (P) (KREEP) are confined in the western maria. The absolute abundances of Th in the highlands had been significant for the lunar evolution. The AGRS results also showed that there was a compositional anomaly on the farside, initially called the big backside basin, now called South Pole–Aitken (SPA).

The map in Fig. 5.1(b) shows the Th abundance on the Moon. The data are only available for portions of the Moon overflown by Apollo 15 and 16. Red, yellow, green, blue, and purple indicate the order of the abundances, where red indicates the highest abundance. Mare regions on the western part of the lunar nearside show relatively high Th abundance, while mare regions on the eastern part of the lunar nearside have lower Th abundance. The lowest Th abundance is on the lunar farside (BVSP, 1981; Heiken *et al.*, 1991).

5.1.2 *Apollo X-ray fluorescence spectrometer*

Passive remote-sensing X-rays have been used to obtain elemental composition maps of planetary bodies (Adler *et al.*, 1972a). The first measurements were carried out during the Apollo 15 and 16 missions (Adler *et al.*, 1972a; 1972b). The Apollo measurements were carried out during the solar minimum and only data for the relative composition of Mg, Al, and Si were obtained. The Apollo X-ray fluorescence spectrometer (AXFS) consists of the following three main subsystems: (1) three large area-proportional counters with 0.0025-cm-thick beryllium (Be) windows; (2) a set of large area filters for energy discrimination for the two proportional counters, thus the three proportional counters have different sensitivities to X-ray energy, which distinguish the characteristic X-rays of Al, Si, and Mg; and (3) a data handling electronics system.

The detector assembly senses X-rays emitted from the lunar surface and converts them into voltage pulses proportional to their energies which

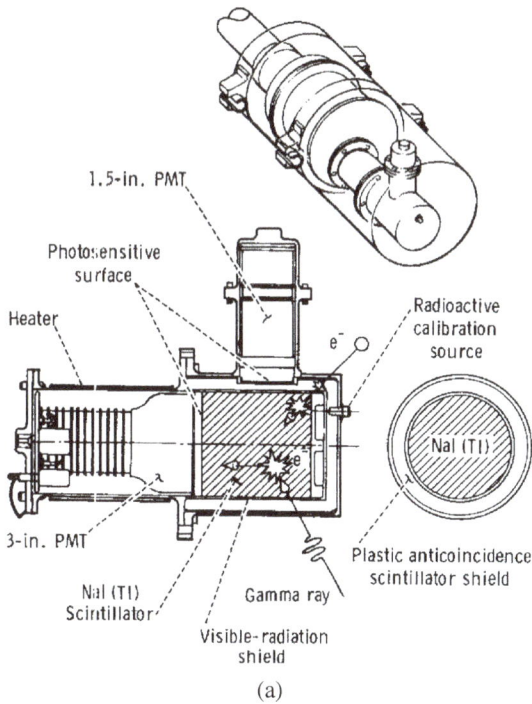

(a)

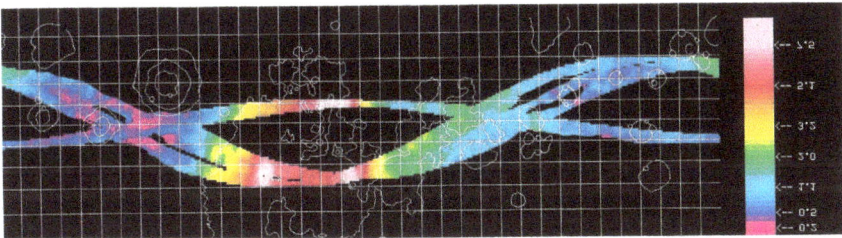

(b)

Figure 5.1:   (a) (top) Schematic drawing of the AGRS (Arnold *et al.*, "16. Gamma-Ray Spectrometer Experiment", APOLLO 15 Preliminary Science Report, NASA SP-289, 1972), (b) (bottom) the Th abundance (ppm) measured by AGRS 15/16 (Metzger, 1993)

are processed in the X-ray processor assembly. The three proportional counters are identical, each having an effective window area of approximately 25 cm². The proportional counters are filled to a pressure of 1 atm with the standard P10 mixture of 90% Ar, 9.5% CO_2, and 0.5% He. To change the energy response, filters are mounted across the Be window aperture on two of the proportional counters. The filters consist of a Mg foil and an Al foil of 5.08×10^{-4} to 1.27×10^{-3} cm thickness. The third counter does not contain a filter. A single-collimator assembly was used to define the field of view (FOV) of the three proportional counters as a single unit. The collimator consists of multicellular baffles that combine a large sensitive area with high resolution but are restricted in FOV. The FOV determines the total flux recorded from the lunar surface and the spatial resolution.

5.2 Lunar Prospector Gamma-ray and Neutron Spectrometers

5.2.1 *Gamma-ray spectrometer*

The Lunar Prospector Gamma Ray Spectrometer (LPGRS) performed the first "direct and global" analysis of the chemical composition of the entire lunar surface (Feldman *et al.*, 1999; 2004a). The LPGRS consisted of a 7.1 cm in diameter × 7.6 cm in length cylinder of bismuth germinate (BGO) placed within a well-typed borated plastic scintillator (BC454) anticoincidence shield (ACS) as shown in Fig. 5.2(a) (Feldman *et al.*, 1999; 2004a). The ACS is of 12 cm in diameter × 20 cm in length and contains a 10-cm deep by 8.4-cm wide well-shaped cutout that houses

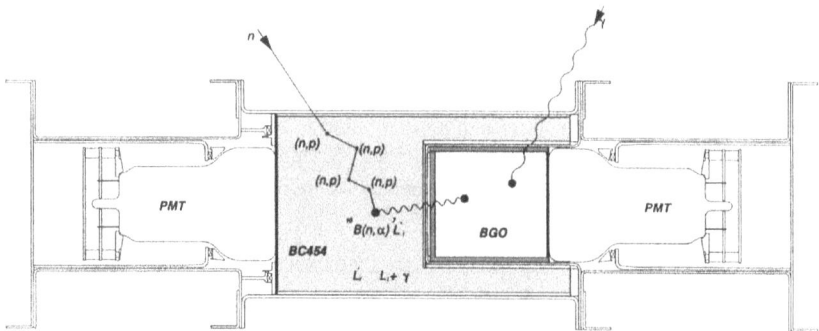

Figure 5.2(a)   Schematic drawing of the gamma-ray spectrometer (GRS) of Lunar Prospector (Feldman *et al.*, 1999; 2004a).

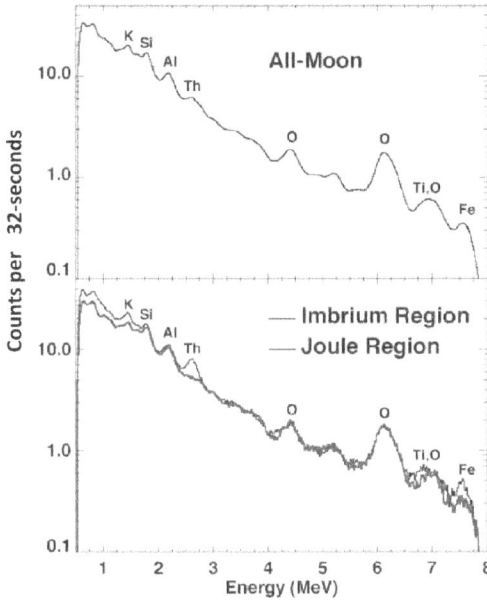

Figure 5.2(b): Lunar Prospector spectra plotted as number of counts per 32 s versus energy. The top panel shows an average gamma-ray spectra collected during the first 5 months of the mission. The bottom panel shows spectra taken for two regions, the Mare Imbrium and Joule Regions enriched with **KREEP** and anorthosite, respectively (Lawrence *et al.*, 1998).

the BGO crystal on one end. The 512-channel gamma-ray spectra, both accepted and rejected by the ACS, were recorded separately and telemetered to Earth. The GRS was mounted on the end of one of the three 2.5-m radial booms extending from the Lunar Prospector. The orbital altitude was approximately 100 km with a surface resolution of ~150 km. The GRS spectra for the entire Moon and two specific regions are shown in Fig. 5.2(b). The Th abundance was obtained from the counting rate of the 2.6 MeV line produced by the radioactive decay of thorium nuclei during the period from 16 January 1998 to 29 October 1998 (Lawrence *et al.*, 1998).

5.2.2 *Neutron spectrometer*

The Lunar Prospector Neutron Spectrometer (LPNS) consists of a pair of ^{3}He gas proportional counters (Fig. 5.2(c)). They have active volumes

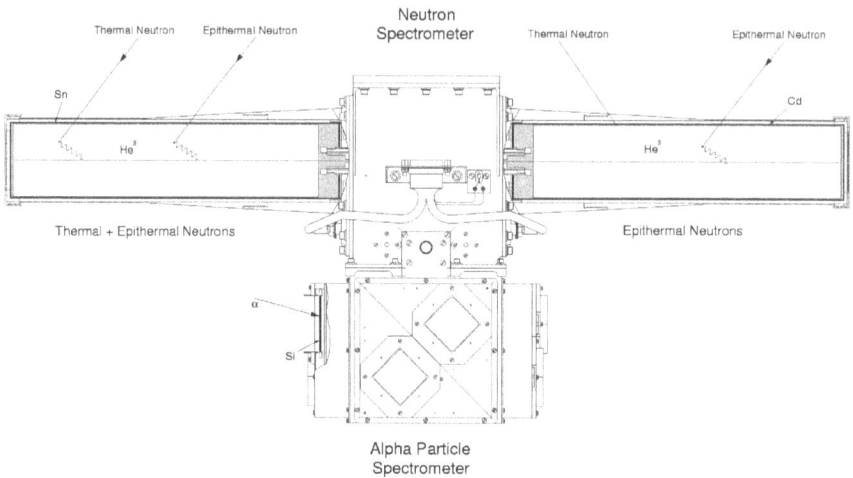

Figure 5.2(c):   Schematic drawing of the two neutron spectrometer ^{3}He gas proportional counters (Feldman *et al.*, 2004a).

with 5.7 cm in diameter and 20 cm in length, and are pressurized to 10 atm (Feldman *et al.*, 1999; 2004a). One of the counters is covered by a 0.63-mm-thick sheet of Cd and so responds only to epithermal neutrons, while the other is covered by a 0.63-mm-thick sheet of Sn, and so responds to both thermal and epithermal neutrons. The difference in their counting rates gives a measure of the thermal neutron flux. These counting rates are measured by window discriminators that cover the 763 keV peaks in the pulse-height spectra.

Feldman *et al.* (2001) found that the hydrogen abundance ranged from 50 to 160 ppm. Lawrence *et al.* (2013) showed that the averaged hydrogen abundance near the poles ranges from 100 to 150 ppm. Furthermore, modeled and measured thermal neutron counting rates suggest that this hydrogen is buried under 10 ± 5 cm of dry soil, which is consistent with the previous determination of a 5-cm burial depth obtained from fast neutrons.

5.3 Kaguya Gamma-Ray Spectrometer

The **Kaguya** Gamma Ray Spectrometer (KGRS) consists of three major subsystems, a gamma-ray detector (GRD), a cooler driving unit (CDU), and gamma-ray and particle detectors electronics (GPE) (Hasebe *et al.*, 2008; 2009). The GRD is mounted on the lunar side of the surface of the spacecraft while the CDU and GPE are installed inside. The GPE contains analog and digital circuit boards for data processing and analysis, and a CPU board for command analysis and data handling and transmission. The GPE controls the CDU, which supplies a cryocooler with a power of 55 W at 52 Hz for the refrigeration of the germanium crystal.

The GRD consists of three radiation detectors, a large High-purity Geranium (HPGe) detector, a massive bismuth geranium oxide (BGO) detector, and a thin plastic scintillator. The HPGe detector, the main gamma-ray detector in the spectrometer system, has an *n*-type, coaxial cylindrical germanium crystal of 65 mm in diameter and 77 mm in length with a volume of ~250 cm^3. The axis of the crystal is placed parallel to the spacecraft surface so that the background gamma rays from the satellite can be minimized by the BGO detector. The HPGe has successfully achieved a superior energy resolution of ~3.0 keV in Full Width at Half Maximum (FWHM) for ^{60}Co 1.33 MeV gamma rays in preflight tests.

Surrounding the HPGe is the BGO detector, which is horseshoe-shaped and has a volume of ~1250 cm^3. The massive and high-atomic-number BGO detector with the maximum thickness of 4 cm, while passively shielding the background gamma rays from the spacecraft, serves as an active anticoincidence detector to suppress Compton-scattered gamma-ray events and cosmic-ray events in the Ge detector. The BGO crystal has the optical reflector and an opening for the Ge detector. The curved plastic scintillator is 5 mm in thickness, and placed on the lunar side of the Ge detector. The scintillator is transparent to gamma rays and rejects charged particle events. The cross section of the scintillator is shown in Fig. 5.3(a).

The energy spectrum of gamma rays acquired by the GRD is shown in Fig. 5.3(b). The high-precision KGRS was carried on Japan's first large-scale lunar explorer, SELENE (Kaguya), which was successfully launched by the H-IIA rocket on 14 September 2007. Energy spectra including many clear peaks of major elements and trace elements on the lunar surface have been measured by the KGRS (Hasebe *et al.*, 2009).

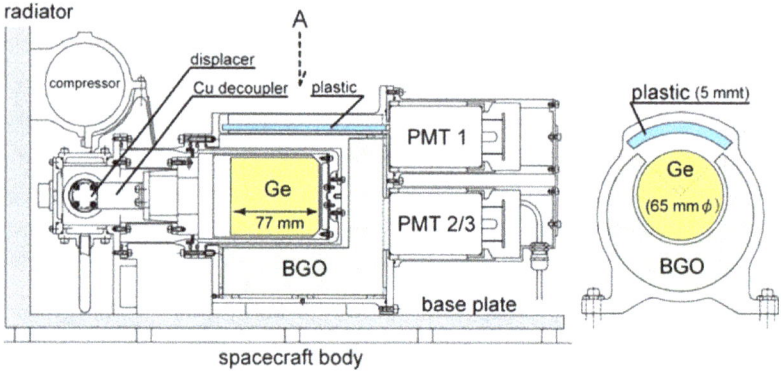

Figure 5.3(a):   Schematic drawing of the gamma-ray spectrometer on Kaguya (Hasebe *et al.*, 2008).

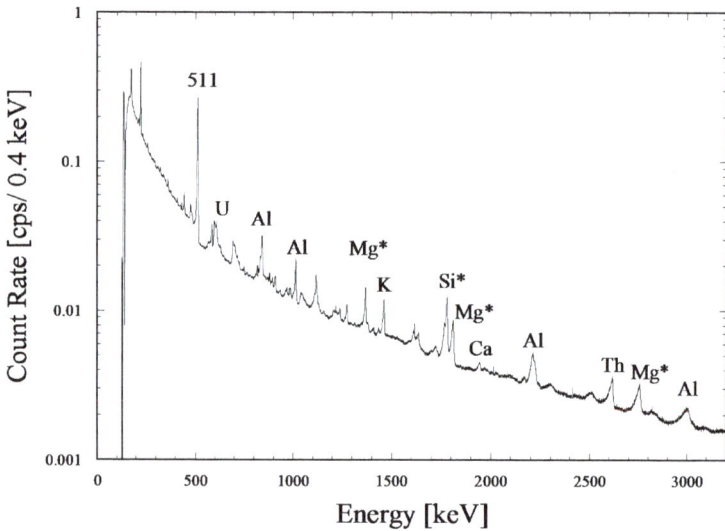

Figure 5.3(b):   Energy spectrum of gamma rays with energies up to 3 MeV measured by KGRS (Hasebe *et al.*, 2009).

5.4 Chang'E-2 Gamma-Ray Spectrometers

The main detector of Chang'E-2 gamma-ray spectrometers (CGRS2) is the redesigned Chang'E-1 GRS (Ma *et al.*, 2008; 2013). $LaBr_3(Ce)$ was used as the main detector instead of CsI(Tl) in Chang'E-1 GRS. Its energy resolution is significantly better than that of Chang'E-1 GRS. The handling of LaBr$_3$ is easier compared to a HPGe detector, which requires cooling. The main drawback of LaBr$_3$, however, is its high intrinsic background gamma rays. The background counting rates due to the radioactivity of LaBr$_3$ itself can be reduced by subtracting the gamma ray spectrum obtained in a low-background chamber (Ma *et al.*, 2013). The CGRS2 employed a LaBr$_3$(Ce) scintillator of 10.8 cm in diameter and 7.8 cm in length as the main detector and an energy resolution of 3.6% at 662 keV was obtained in the laboratory. The schematic drawing of CGRS2 is shown in Fig. 5.4(a). The anticoincidence crystal is made of CsI. The GRS consists of two parts, GRD and an electronics control box (ECB). The GRD includes all the scintillation detectors and PMTs. The high-voltage electronics box and preamplifiers are also installed on the GRD. The data processing unit in the ECB provides the digital interface, power supply, and control.

The gamma-ray spectrum was accumulated every 3 s. The energy interval of the spectrum is about 20 keV and 512 channels were recorded. Figure 5.4(b) shows the spectrum from the Moon.

5.5 Chandrayaan-1 High-Energy X-Ray/Gamma-Ray Spectrometer

A high-energy X-ray/gamma-ray (30–270 keV) spectrometer (HEX) was carried on the first Indian lunar mission, Chandrayaan-1 (Goswami *et al.*,

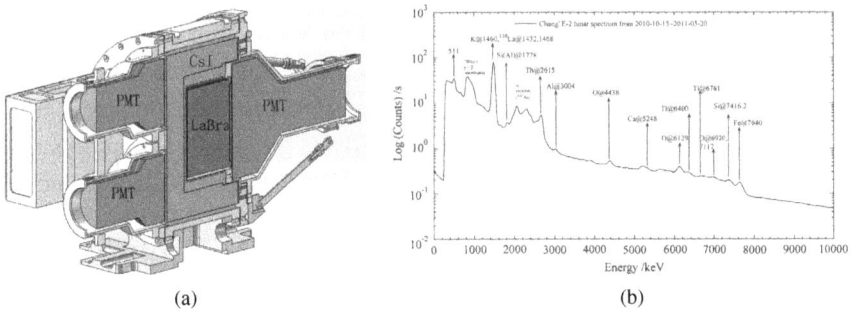

(a) (b)

Figure 5.4(a) (left): The structure of the CGRS2 (Ma *et al.*, 2013) and (b) (right) energy spectrum of lunar gamma rays (Ma *et al.*, 2013).

2005; Vadawale *et al.*, 2014). The Chandrayaan-1 mission was launched on 22 October 2008. The primary science objective of the mission was to study the transport of volatiles on the lunar surface by detecting the 46.5 keV line from radioactive ^{210}Pb, a decay product of the gaseous ^{222}Rn, both of which are members of the ^{238}U decay series.

The HEX employed pixelated Cadmium–Zinc–Telluride (CZT) array detectors. It has a suitable collimator system providing an effective spatial resolution of 40 km in the low-energy region (<60 keV). The system includes a CsI anticoincidence system for reducing the background (Goswami and Annadurai, 2008, 2009).

The Compact imaging X-ray spectrometer (CIXS), which was developed for the ESA's SMART-1 (Grande *et al.*, 2007), was carried on Chandrayaan-1, but these were spatially limited. The new C1XS data set covers approximately 5% of the lunar surface that has not been previously mapped by the XRF. The CIXS, covering 1–10 keV, mapped the abundance of Mg, Al, Si, Ca, Ti, and Fe on the lunar surface with a spatial resolution of 25 km and monitored the solar flux. This experiment utilized planetary X-ray fluorescence spectroscopy to measure the abundances of major rock-forming elements on the lunar surface.

5.6 Mars Nuclear Spectrometer

5.6.1 *Mars Odyssey gamma-ray and neutron spectrometer*

The Mars Odyssey was launched on 7 April 2001 and arrived at Mars on 24 October 2001. The Mars Odyssey has spent more than 10 (earth) years in orbit around Mars, collecting data on Mars' climate and geology. It also served as a key communication relay for the missions to Mars over a period of five years. The Mars Odyssey's gamma-ray spectrometer instrument suite (MOGRSIS) were designed for observation of gamma rays and neutrons from the Mars (Boynton *et al.*, 2004). The MOGRSIS conducted successfully determined 20 chemical elements on the Martian surface, including hydrogen in the shallow subsurface of Mars. The MOGRSIS consists of the following four units: the gamma ray spectrometer (GRS), the neutron spectrometer (NS), the high-energy neutron detector (HEND), and the central electronics box (CEB). The CEB houses the electronics for the gamma and neutron subsystems (GS + NS) and various interfaces, power distribution, housekeeping and command, and data-handling electronics. The GRS weighs 30.5 kg and uses 32 W. Along with its cooler, the GRS measures 46.8 cm × 53.4 cm × 60.4 cm in size.

GRS: The major components of the GRS assembly are the HPGe, the two-stage cooler subsystem, the door, and the gamma pulse amplifier as shown in Fig. 5.5(a) (Boynton *et al.*, 2004). The semiconductor detector is a large n-type HPGe of about 6.7 cm in diameter and 6.7 cm in length. The GRS is separated from the spacecraft (S/C) by a 6 m boom, which was extended several months after the S/C entered the mapping orbit at Mars in order to minimize the S/C background gamma rays. The HPGe detector is at a potential of ~3000 V and a leakage current of less than 1 nA. The detector must be operated at colder than ~140 K, to maintain high resolution and a low leakage current. The GRS has determined reliable elemental abundance maps for H, Si, Fe, Cl, K, and Th (Boynton *et al.*, 2004).

NS: The Mars Odyssey NS detector, as shown in Fig. 5.5(b), consists of a cubical block of boron-loaded plastic (BLP) scintillator (Boynton *et al.*, 2004). The four-segmented units are optically isolated from one another, and each is viewed by a separate 3.8-cm PMT. Both the scintillator assemblies are covered with a 0.69-mm-thick Cd sheet. Global distributions of

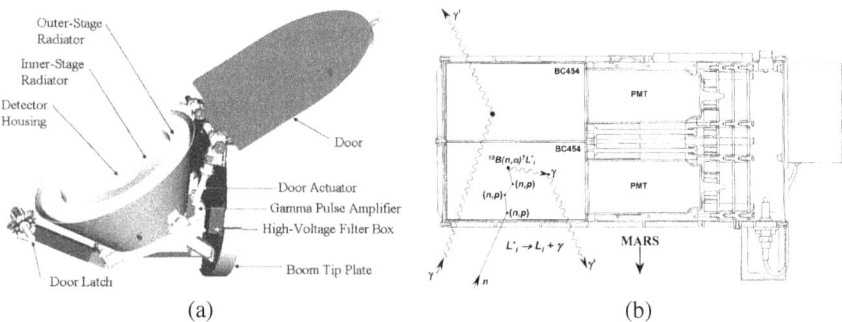

Figure 5.5: (a) (left) Drawing of the GRS gamma sensor head. (b) (right) Drawing of the neutron spectrometer (NS) showing two of the four BC454 prisms in cross section with their associated PMT. A schematic of the interaction of neutrons and gamma rays with the BLP is also indicated (Boynton *et al.*, 2004).

thermal, epithermal, and fast neutron fluxes were obtained during Mars' late southern summer/northern winter using the NS (Feldman *et al.*, 2002).

HEND: The HEND integrates in one unit, a set of five different sensors and electronics boards. The set of three detectors consists of ^{3}He proportional counters and two scintillator counters (Boynton *et al.*, 2004). The small detector (SD) is sensitive for neutrons at the energy range of 0.4 eV–1 keV; the middle detector (MD) is sensitive at the energy range of 0.4 eV–100 keV; and the large detector (LD) is sensitive for neutrons at the energy range of 10 eV–1 MeV.

The HEND found deficits of high-energy neutrons in the southern highlands and the northern lowlands of Mars (Mitrofanov *et al.*, 2002).

5.6.2 *Mars Science Laboratory*

A Mars rover, Curiosity, was carried by the Mars Science Laboratory mission and landed on the Mars at 2012. One of Curiosity's science payloads is an alpha-particle X-ray spectrometer (APXS) for *in situ* determination of rock and soil chemistry (sources). The various payloads for remote and *in situ* measurements acquire samples of rock, soil, and atmosphere, and analyze them in onboard analytical instruments and observe the environment around the rover.

The main objective of the APXS instrument is to characterize the geological context of the rover's surroundings and investigate the processes that formed the rocks and soils. The APXS signal stems from the topmost micrometers for low Z elements such as Na to Si and some 20 s of micrometers for heavier elements like Fe, mainly determined by the range dependent on the energy of the characteristic elemental X-ray.

The APXS for Curiosity rover, as shown in Fig. 5.5(c), is an improved version of the APXS that flew successfully on Pathfinder and the Mars Exploration Rovers (MER) Spirit and Opportunity (Gellert *et al.*, 2006; Rieder *et al.*, 1997; 2003). The MSL APXS takes advantage of a combination of the well-established terrestrial standard methods of particle-induced X-ray emission (PIXE) and X-ray fluorescence (XRF) to determine elemental chemistry. It uses ^{244}Cm sources ($T_{1/2}$ = 18 years) for X-ray spectroscopy. The APXS is mounted on the robotic arm in the rover's

Figure 5.5(c):   The APXS used on the Mars Exploration Rovers. Image credit: NASA/ JPL-Caltech/Cornell/Max Planck Institut für Chemie/University of Guelph.

body. The sensor head of the silicon drift detector (SDD) was typically placed less than 2 cm away from the sample.

The MSL APXS improved the sensitivity of the MER APXS. A full analysis with detection limits of 100 ppm for Ni and ~20 ppm for Br requires 3 hours, while a "quick look" analysis for major and minor elements at about 0.5% abundance, such as Na, Mg, Al, Si, Ca, Fe, or S, can be done in 10 min or less.

5.7 MESSENGER Gamma-Ray and Neutron Spectrometer

MESSENGER (MErcury Surface, Space ENvironment, GEochemistry, and Ranging) was successfully launched in 2004, and journeyed for more than six years before entering orbit around Mercury. MESSENGER's instrument (MGNS), consisting of a gamma-ray spectrometer (GRS) and neutron spectrometer (NS) (Goldsten *et al.*, 2007) (see Fig. 5.6(a) and 6(b)), yielded maps of the elemental composition of Mercury's surface. The GRS detects gamma-ray emissions in the energy range of 0.1– 10 MeV. The encapsulated crystal was maintained at cryogenic temperatures (90 K) despite the intense thermal environment. The GRS sensor is mounted on the body of the spacecraft, and the outer housing of the GRS crystal is equipped with an anticoincidence shields (ACS) to reduce background from charged particles. The NS consists of a sandwich of three scintillation detectors designed to measure the flux of ejected neutrons in three energy ranges from thermal to ~7 MeV. The NS is particularly sensitive to H content.

GRS: The GRS detector is a high-resolution n-type coaxial HPGe crystal of 50 mm in diameter and 50 mm in length. The HPGe crystal is rigidly clamped in a hermetically sealed Al capsule pressurized with clean and dry nitrogen. The capsule is cooled to the 85–95 K range by a mechanical cryocooler attached to an external passive radiator that rejects its heat into cold space. A plastic scintillator ACS (coupled to a PMT) surrounds the Ge detector, for the rejection of background from galactic cosmic rays (GCRs).

NS: The NS consists of three scintillators, each wrapped separately in ultraviolet-reflective materials and optically coupled to a separate PMT. A 100 mm cube of BC454 scintillator is sandwiched between two 100 mm square scintillators of Li-glass with thickness of 4 mm. The Li-glass scintillators are primarily sensitive to thermal neutrons. The open BC454 faces are covered by a 0.25-mm-thick sheet of Gd. Then, the BC454 detector is sensitive to epithermal neutrons between 0.5 eV and 0.5 MeV and fast neutrons between 0.5 MeV and 7 MeV deposited energy.

5.8 NEAR X-Ray/Gamma-Ray Spectrometer

The Near Earth Asteroid Rendezvous (NEAR) mission was launched on 17 February 1996. NEAR is the first spacecraft to orbit an asteroid, under NASA's Discovery Program. The NEAR's major objectives were to answer fundamental questions about the nature and origin of near-Earth objects. The NEAR spacecraft took ~3 years to reach 433 Eros, one of the largest near-Earth asteroids.

The X-ray/gamma-ray spectrometer (XGRS) on the NEAR spacecraft (Figs. 5.7(a) and (b)) is the primary instrument for determining the surface composition of Eros. The XGRS consists of two state-of-the-art sensors: an X-ray fluorescence spectrometer and a gamma-ray spectrometer (Trombka *et al.*, 2001).

The X-ray fluorescence detector contained three gas-filled proportional counters observing X-ray line emissions from the asteroid. Balanced

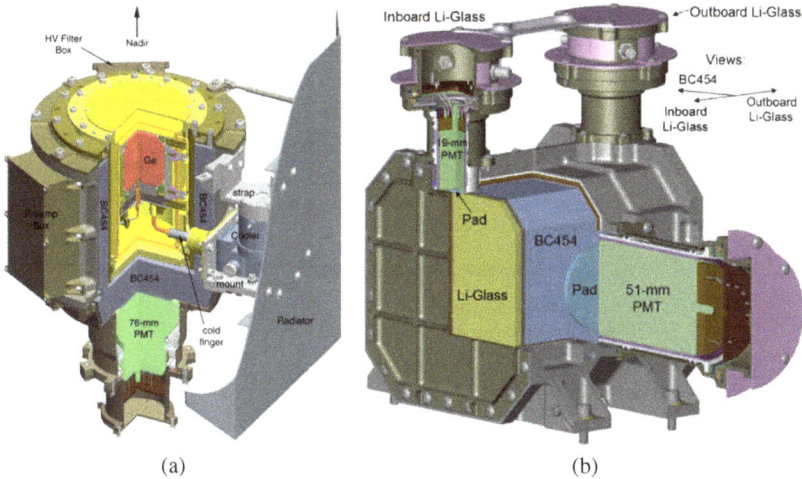

Figure 5.6:   (a) (left) Cutaway view of the GRS sensor on MESSENGER. The upper left part of the passive radiator was cut away to show the internal parts of the instruments. (b) (right) Cutaway view of NS sensor on MESSENGER. (Goldsten *et al.*, 2007).

filters on two detectors (Al on one and Mg on the other) were used to separate Mg, Al, and Si lines; Ca, Ti, and Fe lines were also resolved. The solar monitor uses an additional gas-filled proportional counter with a pinhole active area observing the solar X-ray spectrum.

The gamma-ray spectrometer contained a body-mounted NaI scintillator (2.5 cm in diameter and 7.5 cm in length) as a main detector with a BGO shield (8.9 cm in diameter and 14 cm in length). The energy range of gamma rays covers 0.1–10 MeV in 1024 channels.

5.9 Dawn Mission Gamma-Ray and Neutron Spectrometer

The Dawn spacecraft of a NASA Discovery class mission explores two of the largest main-belt asteroids, 4 Vesta and Ceres whose accretion is believed to have been interrupted by the early formation of Jupiter. A gamma-ray and neutron spectrometer (GNS) determines the elemental composition of these protoplanetary bodies, providing a better understanding of processes occurring shortly after the onset of the condensation of the solar nebula (Prettyman *et al.*, 2011). GNS contains four types of radiation spectrometers (see Fig. 5.8).

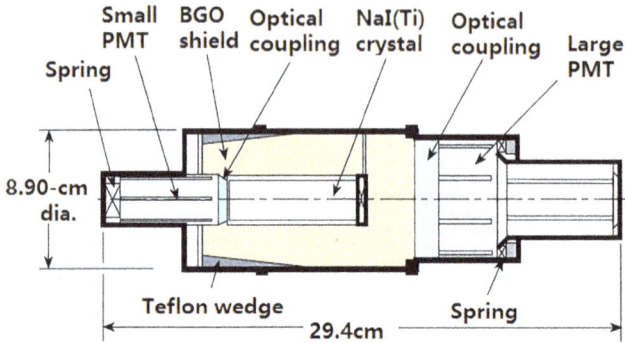

Figure 5.7(a): Schematic drawing of the NEAR gamma-ray spectrometer (Trombka, *et al.*, 2001).

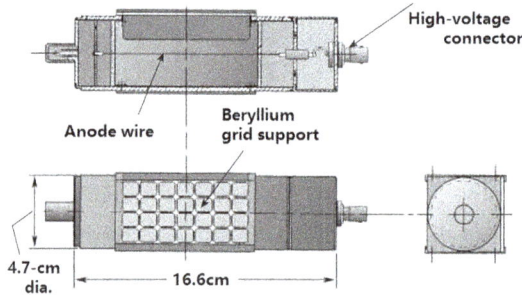

Figure 5.7(b): Schematic drawing of the NEAR gamma-ray spectrometer (Trombka, *et al.*, 2001).

The spectrometer includes the following:

BGO scintillator: A 7.6 cm × 7.6 cm × 5.08 cm BGO (~300 cm³) is located at the center of the sensor head. The scintillator is read out by a 51-mm PMT (Hamamatsu: R6231). The BGO has high sensitivity to gamma rays over a wide energy range (up to 10 MeV), though the energy resolution at room temperature is ~10% in FWHM at 662 keV.

CZT semiconductor: A planar array of 4 × 4 CZT crystals is positioned on the +Z side of the BGO (see Fig. 5.8) (Prettyman *et al.*, 2003; 2011). Each crystal is 10 mm × 10 mm × 7 mm, and mounted in a ceramic package. The array has an active volume of 11.2 cm³. Coplanar grids (CPG) in the

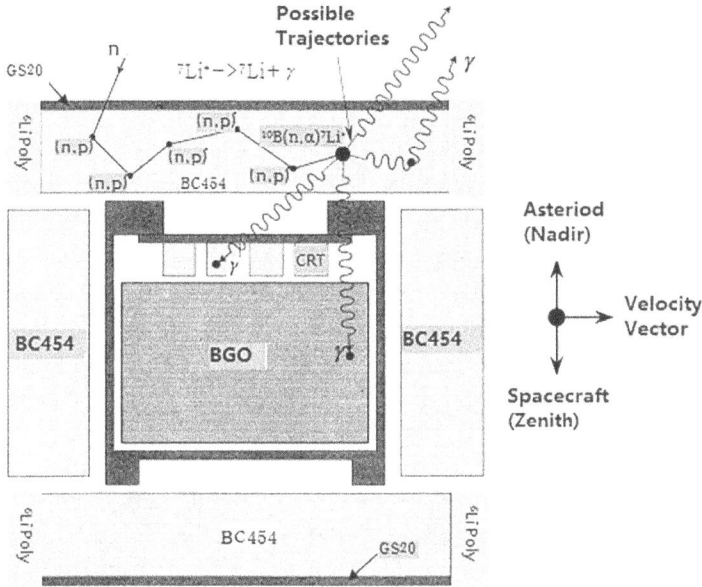

Figure 5.8:   Cross-sectional view of GNS showing the arrangement of neutron and gamma-ray-sensitive elements (Prettyman *et al.*, 2003).

CZT read-out electrodes were used for charge collection and sensing to mitigate the effects of hole trapping, resulting in excellent energy resolution over a wide range of energies (Luke, 1995). The energy resolution was better than 3% FWHM at 662 keV.

BLP scintillator: Two L-shaped BLP scintillators (each 193 cm^3) are located on the $+/-$ *Y* sides of GNS, surrounding the sides of the BGO crystal and CZT array. All the BLP scintillators on GNS are EJ254 (Eljen Technology, containing 5% natural B, identical to BC454). The scintillators are viewed by 1-inch PMTs (R1924A-01). The BLPs act as an ACS to reject GCR interactions. The scintillators are also sensitive to neutrons. Fast neutrons (>700 keV) can undergo elastic scattering with hydrogen atoms.

Phosphor sandwich (phoswich): Two BLPs are located on the nadir (+Z) and spacecraft (−Z) sides of the instrument. Each BLP is ~10.16 cm × 10.16 cm × 2.54 cm and is read out by a 1″ PMT (Hamamatsu, R1924A-01). Each scintillator is covered with a sheet of Gd foil. The outward facing side is covered by a 0.2-cm-thick lithiated glass (LIG). The LIG is optically coupled to the BLP such that the PMT measures light produced in both the glass and the plastic.

5.10 Rosetta Alpha Particle X-Ray Spectrometer

Rosetta is a cornerstone mission to land on a comet. The Rosetta mission, with a combination of remote sensing and *in situ* measurements, was launched in 2004 with the main objective to gain a better understanding of the origin and formation of comets and the solar system. As a part of the lander's payloads, the Rosetta APXS (RAXPS) measured the chemical composition of the comet's surface *in situ* (Klingelhoefer *et al.*, 2007).

RAPXS: The arrangement of the Rosetta detectors and alpha sources is strictly concentric (see Fig. 5.9). The X-ray detector is in the center of the front side. Six alpha-ray sources (~30 mCi of ^{244}Cm in total) are placed in a circle around the center of the sensor head on a movable holding device. Six alpha-ray detectors are arranged concentrically around the alpha-ray sources. The alpha-ray detectors have a thickness of 300 μm. The sensor head is located on the outside of the lander. The nominal working distance between the detectors and the sample is ~30 mm. The backside of the doors made of Cu–Be alloy, $Cu_{98}Be_2$, serves as a calibration target.

This craft was launched in the year 2004 with the aim to observe the detailed structure of the comet named 67P/Churyumov–Gerasimenko (Hand, 2015).

Carboxylic acids, precursors of amino acids, were observed on this comet, as well as the existence of various organic molecules (Capaccioni *et al.*, 2015). The detected organic molecules form only in extremely cold environments as a result of UV light and cosmic rays striking various types of ice-coated dust. Philae, Rosetta's lander, was released to land on the comet to observe surface material, including icy matter. However, the locations of the landing of this lander have not yet been clarified. So, it is necessary for us to wait patiently until the locations and any other observed results are available to this mission's researchers.

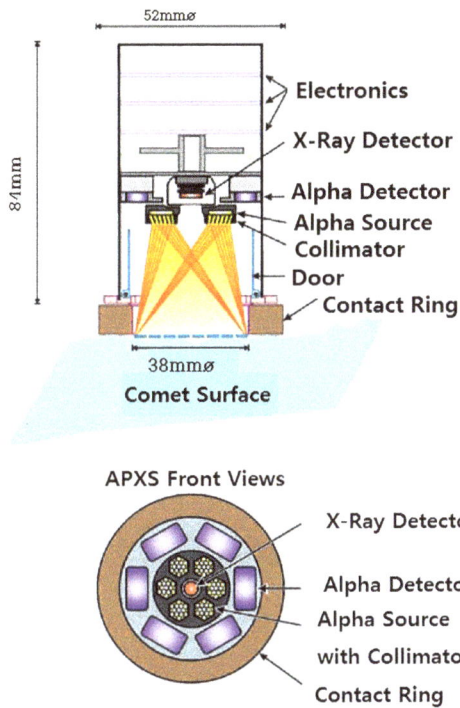

Figure 5.9:   Schematic drawing of the APXS sensor head in measurement position (Klingelhoefer *et al.*, 2007).

Chapter 6

Data Reduction

From the data observed at spacecraft altitude, we would like to obtain the properties of material at the ground. Here, we focus on the derivation of elemental composition from the gamma-ray spectrometer (GRS) data. The GRS data reduction depends on the type of detector system, because the transmission of the data preserved spatial resolution and permitted complete flexibility in binning the data according to any desired arrangement of time or location in space. Translating the accumulated pulse-height spectrum into elemental concentration is complex, requiring knowledge of the relevant background components, the response of the instrument, the observing geometry, and the production of gamma rays per unit surface concentration (Pieters and Englert, 1993). For low-resolution detectors, response function analysis and energy band analysis are used. For example, NaI(Tl) and other scintillator detectors broaden the incident gamma-ray lines; the continuum dominates the spectrum, contributing about 85% of the response, the precise proportion varying with energy and surface composition (Pieters and Englert, 1993). For high-resolution detectors, photopeak analysis is used.

6.1 Response Function Analysis

The response of a detector to an incident gamma ray is dictated by three interaction mechanisms: photopeak absorption, Compton scattering, and pair production, whose relative effectiveness depends on the size, shape, and material of the detector as well as the energy of the gamma ray. Individual response functions correspond to monenergetic gamma rays as shown by the pulse-height spectra formed due to 0.6616-MeV gamma rays from ^{137}Cs for the Ge detector in Fig. 3.2 and for the NaI(Tl) scintillator in Fig. 3.4. A set or library of response functions specific to the low-resolution detector needs to be obtained using a standard gamma-ray source and Monte Carlo calculation. The emissions of one to several gamma rays may contribute to significant emissions from a particular element. Individual response functions corresponding to monoenergetic gamma rays can be added in proportion to their relative intensity to produce a spectrum characteristic of emission from a particular element (Trombka *et al.*, 1979). This analysis is involved with the complexity of the composite discrete spectrum and the use of the entire monoelemental response function to characterize a particular element rather than confining the analysis to the area of the photopeak. This technique uses a least-squares fit of selected library response functions covering the entire energy range of the measured data. Both the Apollo 16 and Lunar Prospector GRSs are associated with low-resolution detectors such as NaI(Tl) and BGO, respectively. The data reduction of these detectors was accomplished using this technique. The library of monoelemental response functions is fitted to the discrete line spectrum by means of a matrix inversion analysis. Subtracting the total discrete line spectrum from the original spectrum gives a second estimate of the continuum. The process is repeated until a good fit is obtained for both the continuum and the monoelemental response functions.

6.2 Energy Band Analysis

Energy band analysis is associated with low-resolution detectors. For this method, the abundance of a particular element is determined through variations in count rate over a specific portion of the energy spectrum where that element is dominant. This technique is most widely used for the analysis of thorium (Th) using the energy band from 0.55 to 2.75 MeV. In the energy region above the positron line at 0.511 MeV, up to and including the highest energy line due to natural radioactivity, namely the 2.61 MeV line due to Th, the regional differences in count rate are predominantly due to varying concentrations of the radioactive elements Th, uranium (U), and potassium (K). Since the statistical precision of the total count rate in this region is excellent, the best possible areal resolution can be obtained. This technique analyzes a particular element through variations in count rate over specific portions of the energy spectrum where its effect is greatest (Pieters and Englert, 1993). Composition can be obtained by matching a set of net GRS count rates with known surface concentrations. A regression line is defined by the relationship. The perturbing effect of other elements on the band count rates must be evaluated and removed if significant. This technique uses the integrated count rate over limited energy intervals, does not require the separation of the GRS spectrum into line and continuum component, and yields results for one element at a time. This technique can also be used for shorter periods of observation (Pieters and Englert, 1993).

6.3 Photopeak Analysis

A good high-resolution detector not only provides a much better energy resolution; it also allows peaks in GRS spectra to have a peak-to-Compton ratio greater than 50, which is about a factor of 10 greater than that for an NaI(Tl) detector (Pieters and Englert, 1993). Equipped with such a detector, many peaks are identified and precisely characterized in the spectrum. The Mars Odyssey GRS and SELENE (Kaguya) GRS detectors, both with a high-purity Germanium (HPGe) crystal, identified more than 200 peaks (Evans *et al.*, 2002; 2006; Hasebe *et al.*, 2008; Kobayashi *et al.*, 2010). These detectors were calibrated and peak efficiency with respect to energy was investigated. The analysis of photopeaks can be accomplished with peak analysis software. Often the analysis requires little user intervention. This analysis is confined to a limited energy range within which one to several elements of interest have photopeaks or escape peak features. The analytic model used to describe the spectral data includes the contributions of photopeaks and/or escape peaks, the continuum, and a linear background. The GRS detector for Messenger (MErcury Surface, Space ENvironment, GEochemistry, and Ranging) is also made up of a HPGe crystal. This GRS system enables us to measure natural radioactive elements and major elements to understand the surface of Mercury (Goldsten *et al.*, 2007).

6.4 Ground Truth Analysis

In the case of lunar exploration, for example, we have many return samples of rock and regolith from various landing sites of the Moon, brought to Earth by the Apollo and Luna programs. Moreover, it is known that some meteorites originate from the Moon — these are called lunar meteorites. Return samples and lunar meteorites are investigated in the laboratory to determine their elemental composition. By assuming that some samples from the specific site of the Moon represent the elemental abundance of the site, we now have the "ground truth" of the site. From feldspathic lunar meteorites, calcium (Ca) concentration is estimated for feldspathic highland terrane in the farside region of Moon. The vertical axis of Fig. 6.1 represents the abundance of ^{40}Ca at various ground truth sites. The horizontal axis of the figure represents the count rate of 3737-keV gamma rays from ^{40}Ca remotely sensed by the Kaguya gamma-ray spectrometer (KGRS) over the respective ground truth sites (Yamashita *et al.*, 2012).

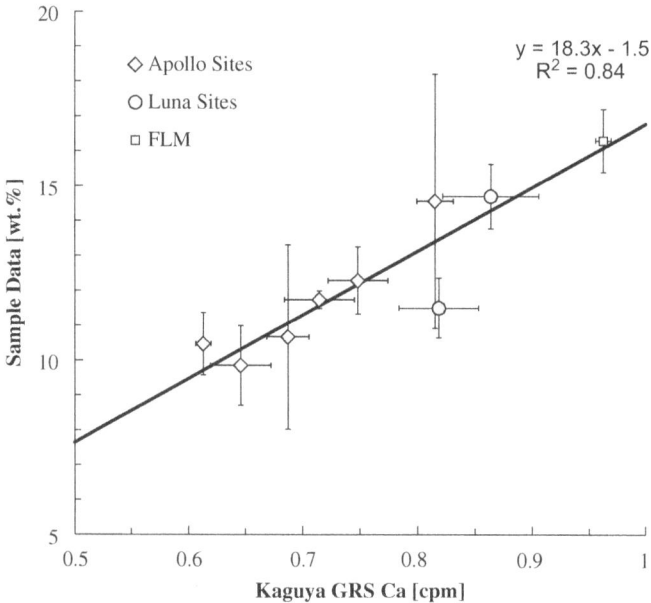

Figure 6.1:   The abundance of Ca in the samples from various sites on the Moon versus observed gamma-ray count rates for ^{40}Ca over the respective sites.

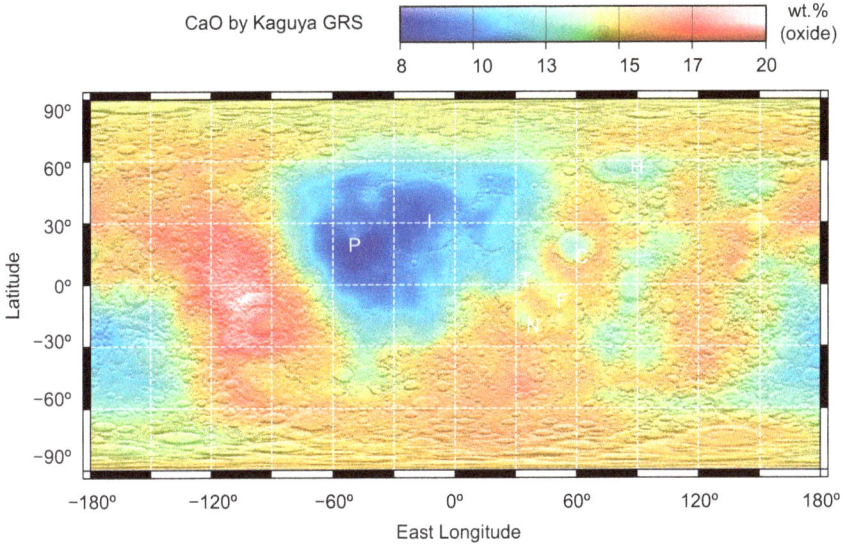

Figure 6.2: Abundance map of Ca derived by count rates of 3737-keV gamma rays from ^{40}Ca (Yamashita *et al.*, 2012).

A linear relation is obtained between the ground truth abundances and the count rates of the ^{40}Ca gamma-ray lines by KGRS. Hereafter, one may obtain the concentration of Ca, y (wt.%), at a lunar site from the count rate, x (cpm), of 3737-keV gamma rays over this site by the relation $y = 18.3x-1.5$. Thus, Yamashita *et al.* (2012) obtained the abundance map of Ca over the whole surface of Moon as shown in Fig. 6.2. It should be noted that this method to derive Ca concentration does not need the flux of fast neutrons though the 3737 KeV gamma ray is emitted by the inelastic scattering of fast neutrons with ^{40}Ca.

6.5 Spatial Deconvolution Analysis

Spatial deconvolution analysis has been used in the GRS data analysis for both Apollo 16 and Lunar Prospector (Lawrence, 2007; Pieters and Englert, 1993). Because of the broad spatial resolution of the GRS data, it is difficult to determine the desired elemental abundance of small-scale geologic features. In the case of the data analysis for Apollo 16, Aristarchus and its surroundings were the first areas to which the spatial deconvolution technique was applied (Haines *et al.*, 1978). Besides Aristarchus, regions centered on Mare Imbrium, the Apennine Mountains, Mare Smythii, and the central highlands have also been deconvolved. Each data field has been modeled into about a dozen constructs of varying size and shape. The end product has been not one model but a family of best fits that are not distinguishable statistically and that show no major variation in model shape. One should note that the improvement in spatial resolution and accuracy is accompanied by a 50–100% loss in precision; both concentration range and uncertainty for a given construct within the family of best fits decrease with each increase in observation time and construct size (Pieters and Englert, 1993).

The detailed process of global spatial deconvolution analysis for Lunar Prospector was described using two spatial deconvolution methods: the Jansson and Pixon methods (Jansson, 1997; Pina and Puetter, 1993). In theory, the Jansson's method is easy to implement in a simple analysis program; but in practice, the iterations are performed to the point where the residual map has the same statistical spread as the noise estimates of the original data. This overfits the data and risks significant amplification of noise (Lawrence *et al.*, 2007). However, the Pixon's deconvolution technique is to obtain the smoothest possible image as constrained by both the original data and the data uncertainty (or noise). This method uses variably sized smooth patches, or Pixon elements, within the image to express the information content of the image. This allows a successful usage in variety of image analysis fields, including astronomy, microscopy, and medicine (Puetter *et al.*, 2005). Spatial deconvolution analysis of Lunar Prospector Th abundances shows a better result with the Pixon method. The histogram of Pixon residuals shows mostly a Gaussian shape with a width (0.96) that is close to 1 and an offset that is less than 1% of the standard deviation; however, the Jansson histogram has a significant negative offset. Both the residual map and the histogram show that the Pixon method gives a much more consistent result with respect to the data uncertainties as compared with that from the Jansson result (Lawrence *et al.*, 2007). However, with these techniques, Th abundance for specific geological features was not determined satisfactorily. Moreover, complex areas require further investigation with an advanced statistical analysis.

6.6 Independent Component Analysis

Forni *et al.* (2009) prepared the spectrum data of gamma rays obtained by KAGUYA GRS on a 3072 equal area over the Moon, and tested the independent component analysis method for analyzing the spectrum data. They identified two main components. The first one was related to radioactive elements U, Th, and K, and the second one was related to iron. They obtained a map of natural radioactive elements U, Th, and K and a map of iron. They compared iron data from the KAGUYA GRS with those of the Lunar Prospector GRS (Lawrence *et al.*, 2002) and found good correlation between the two.

Chapter 7

Major Advances in Nuclear Planetary Science

The scientific results obtained by the nuclear methods for the exploration of the Moon, Mars, Mercury, and some asteroids are mentioned in this chapter. In Section 7.5, future nuclear science missions involving the Moon, Martian satellites, and asteroids are described.

7.1 Moon

Earth's Moon is unique in the solar system in size; compared with that of other planets, our Moon is quite large. The average density of the Moon is 3.312 g/cm^3, which is only about three-fifths of that of Earth. Although lunar surface rocks are essentially similar in matter as the cooled igneous rocks from the mantle of Earth, the interior of the Moon must somehow be different from that of Earth. The Moon has essentially no water and no atmosphere. The dynamic range of temperatures in the equatorial region is extremely wide: from −173°C shortly before lunar dawn to 127°C at midday. Surface temperatures at the lunar poles are much colder, ranging from −258°C to −113°C, due to the very low input of sunlight. The radiation environment on the Moon is quite different from that on Earth in spite of the Moon being the nearest astronomical object to Earth. Because the Moon has an extremely thin atmosphere and low magnetic field, it does not provide protection to its surface from galactic cosmic rays (GCRs), solar particle events, and high-speed micrometeorites.

The highest elevated point of Moon is on the southern rim of the Dirichlet–Jackson Basin (10.75 km), which is on the farside highland, and its lowest point is inside the Antoniadi crater (–9.06 km) situated in the lunar South Pole-Aitken (SPA) basin (Araki *et al.*, 2009). The lunar high-lands are mainly composed of anorthosite that largely includes feldspar mineral. A recent observation result of SELENE (Selenological and Engineering Explorer; also known as Kaguya) showed that anorthosite consisting of nearly 100% anorthite is widely distributed in the lunar highland crust. Conventionally, the lunar highland crust is believed to consist of 90% anorthite and 10% other minerals (Ohtake *et al.*, 2009).

Basins are lunar impact excavations with diameters of at least 300 km, whereas craters are smaller than 300 km. Two important basins are the Imbrium Basin and the SPA basin. The Imbrium Basin is filled with lava and appears as a dark circle on the western part of the lunar nearside. The SPA basin on the farside of the Moon is the largest impact crater in the solar system, with a diameter of 2500 km and a maximum depth of 12–13 km.

The round, dark, low areas on the Moon, called maria, are surrounded by terrane. The lunar basins that contain the maria were formed by giant impacts early in lunar history and are filled with lava by volcanic activity. The visible maria cover 16% of the surface on the lunar nearside. The rocks of maria closely resemble terrestrial basalts, which are a type of dark and fine-grained volcanic rock. Many lunar maria are associated with gravitational anomalies called mascons. The strongest mascon is associated with Mare Imbrium, though mascons also occur with Serenitatis, Crisium, Humorum, Nectaris, and Orientale.

What is interesting and attractive about Moon is that it records a violent past and has a lot of places to explore. And the next step in lunar exploration will be a program of detailed surface explorations and experiments with autonomous and teleoperational robotic explorations before the return to human activity on Moon.

7.1.1 *Lunar crust*

7.1.1.1 *Major crustal provinces*

Studies of both lunar samples and remote sensing data obtained by the Apollo program gave us the broad outline of the nature and geologic history of the Moon. Many known facts and beliefs, however, are now being questioned on the basis of global data obtained by two US lunar missions, Clementine and Lunar Prospector. In the late 2000s, gamma-ray spectrometers (GRSs) on SELENE (Kaguya) and Chang'E space-craft confirmed the previous data and gained new findings. The lunar data obtained in the late 1990s and 2000s are being integrated with new and old lunar sample data, and give us new ideas about the nature of the Moon. As shown in Fig. 7.1, the lunar crust is divided into three dis-

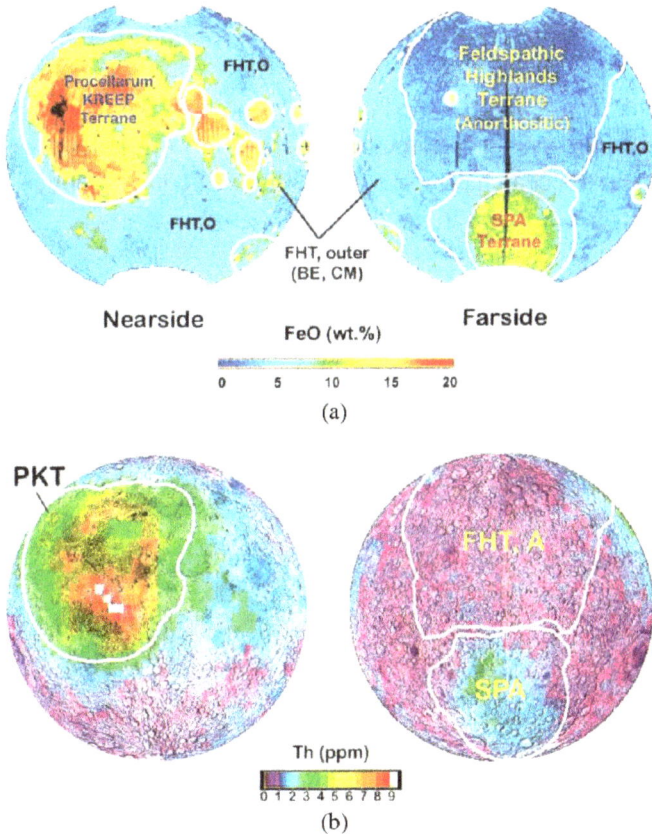

Figure 7.1:   Three major terranes on Moon (Jolliff *et al.*, 2000): (1) the Procellarum KREEP Terrane (PKT), (2) the South Pole-Aitken Terrane (SPAT), and (3) the Feldspathic Highlands Terrane (FHT).

tinct geochemical provinces that are quite different from the traditional highlands and maria: (1) the Procellarum KREEP Terrane (PKT), characterized by high thorium (Th) and iron (Fe) concentrations; (2) the SPA Terrane (SPAT), which has modest iron oxide (FeO) and Th concentrations; and (3) the Feldspathic Highlands Terrane (FHT), which includes its somewhat different outer portion (FHT-O) and has low FeO and Th concentrations (Jolliff *et al.*, 2000).

These prominent regions on the Moon are defined by chemical characteristics — specifically by the concentrations of FeO and Th. Fe and Th, which are particularly useful elements when distinguishing rock types and monitoring geochemical processes. Mare basalts contain lots of Fe and some Th, but feldspar-rich anorthosites do not contain much of them. Th, one of the incompatible elements, behaves like rare earth elements (REEs) that are not incorporated into common minerals. As a result, when magma crystallizes, Th becomes more abundant in the leftover magma; after the magma ocean has mostly crystallized, it contains about 100 times as much Th as that found in the original magma.

7.1.1.1.1 Procellarum KREEP Terrane

Procellarum KREEP Terrane (PKT) is a region that is characterized by high potassium (K), thorium (Th), and uranium (U) concentrations, important components of KREEP (an acronym meaning a lunar chemical component rich in K), rare earth elements (REEs), and phosphorus (P). The PKT is a mixture of assorted rocks, including most of the mare basalts on the Moon, and is characterized by a high Th concentration (about 5 ppm on average). This region has also been called the "High-Th Oval Region" and the "Great Lunar Hot Spot". The PKT was probably formed when the residual, incompatible-element-rich liquid intruded into the primitive feldspathic crust in the last stages of the magma ocean's crystallization (Taylor and Jakes, 1974; Warren and Wasson, 1979).

The Lunar Prospector data demonstrated that the surface abundance of incompatible elements (Th, K, Gd, Sm, and the other elements that make up KREEP) is high in this region, which encompasses that of Oceanus Procellarum, Mare Imbrium, and the adjoining mare and highlands (Elphic *et al.*, 2000; Lawrence *et al.*, 1998; 2000). The results of the

Kaguya GRS (KGRS) are very similar to those of the Lunar Prospector GRS (LPGRS) (Kobayashi *et al.*, 2010; Yamashita *et al.*, 2010).

The Procellarum basalts were found to be enriched with incompatible elements, which shows the unique volcanic history of the PKT on the origin and evolution of this province. The KREEP-rich materials of this region suggest that the final dregs of the lunar magma ocean (LMO) ultimately accumulated within this region. The Imbrium impact event is the last basin forming in the PKT, which is not younger than ~3.84 Gyr (Wieczorek *et al.*, 2006).

7.1.1.1.2 Feldspathic Highlands Terrane

The Feldspathic Highlands Terrane (FHT) is characterized by low FeO (4.2 wt% on average) and very low Th (0.8 ppm) concentrations. This lunar crust is composed of anorthosite and related feldspar-rich rocks that formed about 4460 ± 40 Myr ago (Norman *et al.*, 2003) by flotation from the magma ocean, and represents a pure form of the primary lunar crust (Ohtake *et al.*, 2009). Jolliff *et al.* (2000) subdivided the FHT into FHT-A (anorthositic) and FHT-O, which is somewhat richer in FeO and poorer in Al_2O_3 than FHT-A. The LPGRS data indicate that FHT-A has low Th concentration (0.5 ppm) and 4.3 wt% FeO (Gillis *et al.*, 2004). Because FeO is anticorrelated with Al_2O_3 (Haskin and Warren, 1991), this corresponds to an Al_2O_3 concentration of 26.3 wt%. If anorthosite makes up the bulk of this material, most of FHT-A may have lower Th (0.04 ppm) and higher Al_2O_3 (30–35 wt%) concentrations. Lunar highland meteorites (Korotev *et al.*, 2003) and lunar granulitic breccias (Cushing *et al.*, 1999) may be representative of FHT-O. These rocks typically contain 0.4–1 ppm Th, 3–6 wt% FeO, and 25–28 wt% Al_2O_3.

7.1.1.1.3 South Pole-Aitken basin

The South Pole-Aitken (SPA) basin on the lunar farside is the oldest and most topographically well-preserved lunar impact basin (Spudis *et al.*, 1994; Wieczorek and Phillips, 1999). KREEP appearing low in abundance at this location divided it into an inner zone (SPAT-inner) and an outer one

(SPAT-outer) (Jolliff and Gillis, 2005; Jolliff *et al.*, 2000). SPAT-inner has moderate FeO (average of 10.1 wt%) and Th (1.9 ppm) concentrations whereas SPAT-outer has less FeO (5.7 wt%) and Th (1.0 ppm).

Lucey *et al.* (1998) indicated that this basin is likely to be part of a mixture making up the basin floor based on its FeO and TiO_2 concentrations from the Clementine data. The absolute FeO concentrations were found to be between ~6 and 12 wt% by the LPGRS (Pieters *et al.*, 1997; 2001). The Th abundance of the SPA basin is about ~2–3 ppm, and this is slightly higher than that of the surrounding highlands (~1 ppm).

7.1.1.2 *Crustal composition and thickness*

The Apollo missions returned 381.7 kg of lunar rocks and the Luna missions brought 276 g of lunar regolith to Earth (Hiesinger and Head, 2006). On the basis of their texture and composition, lunar materials can be classified into four distinct groups (1) pristine highland rocks that are primordial rocks, uncontaminated by impact mixing; (2) pristine basaltic volcanic rocks, including lava flows and pyroclastic deposits; (3) polymict clastic breccias, impact melt rocks, and thermally metamorphosed granulitic breccias; and (4) the lunar regolith (Hiesinger and Head, 2006).

Pristine highland rocks can be subdivided into two major groups based on their molar Na/(Na + Ca) content and the molar Mg/(Mg + Fe) content of their bulk rock compositions (Papike *et al.*, 1998). Most highland rocks were formed during the early differentiation of Moon, when the upward separation of buoyant plagioclase within a magma ocean produced a thick anorthositic crust. Mare basalts are enriched in FeO and TiO_2, and olivine and/or pyroxene, are depleted in Al_2O_3, and have higher CaO/Al_2O_3 ratios than highland rocks (Taylor *et al.*, 1991). KREEP basalts differ from mare basalt and were formed by mantle melts of late-stage magma-ocean residua — the so-called urKREEP (Warren and Wasson, 1979).

There are two possible interpretations of the Moon's internal structure: (1) just after magma-ocean crystallization and (2) near the end of mare basaltic volcanism. A model with the hemispheric dichotomy in crustal thickness and mare basaltic volcanism represents the crust and mantle as being laterally uniform in composition.

Seismic data of the lunar surface obtained from the Apollo 12, 14, 15, and 16 stations provide the most direct observation of the crust's thickness. The estimated thickness of the Apollo 12 and 14 landing sites was ~60 km (Nakamura *et al.*, 1982; Toksöz *et al.*, 1972; 1974). However, recent analysis suggested the thinner crust of these regions to be ~27–50 km (Khan and Mosegaard, 2002: Lognonné, 2005). The crust beneath the Apollo 12 and 14 landing sites is found to be thinner than the global average.

The data obtained from the AGRSs onboard Apollo 15 and 16 (Metzger *et al.*, 1977; Trombka *et al.*, 1973) and the LPGRSs (Lawrence *et al.*, 1998) have shown that the surface of the lunar farside is relatively depleted of Th. However, the LPGRS's remote sensing could not specifically determine Th distribution in the farside so far because its Th abundance is very low (<1 ppm) — approximately a tenth or less of that in the PKT (Jolliff *et al.*, 2000). The KGRS (Hasebe *et al.*, 2008) has successfully detected the region with the lowest Th concentration, with the highest energy resolution among the past lunar GRSs (Kobayashi *et al.*, 2010; 2012). As shown in Fig. 7.2, they defined two regions with the lowest Th abundance on the farside, zone A: Dirichlet–Jackson, Korolev, Hertzsprung, and Orientale; and zone B: the western region of Moscoviense and

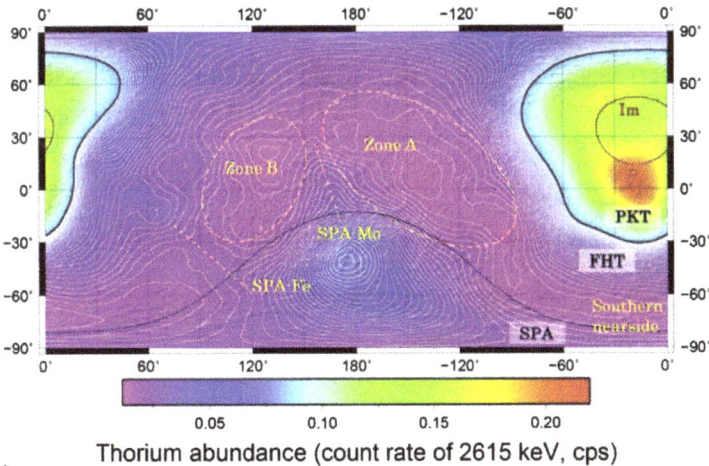

Figure 7.2:   The distribution of gamma-ray intensity of Th measured by the KGRS (Kobayashi *et al.*, 2010; 2012) and the count rate (cps) of ^{208}Tl 2615-keV gamma-ray peak. The shaded relief is produced by using Kaguya-laser altimeter (LALT) data (Araki *et al.*, 2009).

Mendeleev and Tsiolkovsky–Stark. They showed that these regions coincide with the thickest regions of Moon's crust. That is, it shows an inverse correlation between Th abundance and crustal thickness. This fact indicates that a large-scale mechanism of crystallization of the LMO has been involved in the formation of the lunar farside.

The earliest craton crystallized from the LMO would have the lowest Th and the degree of crystallization would be related to Th abundance (Snyder and Taylor, 1993). Th abundance in the surface is inversely correlated to crustal thickness on the farside, and this is consistent with the geochemical property of Th during the crystallization of the LMO (Korotev, 1998).

7.1.1.3 *Petrological composition*

Th concentration as a function of FeO (%) has been investigated for three distinctive rock types — KREEP, feldspathic highland and mare basalt — and volcanic glass (Jolliff *et al.*, 2000). The abundance of Al_2O_3, MgO, TiO_2, CaO, FeO are also used to identify the three types of lunar regolith. Mare rocks contain more FeO, MgO, and TiO_2, but less Al_2O_3 and CaO (Fischer and Pieters, 1996). Surface mixing of KREEP material with volcanic rock is investigated with Th and FeO. In the case of Apollo 14, Th-rich and medium concentrations of FeO (~12%) were observed. In the case of Apollo 16, both Th (~10 ppm) and FeO (<10%) were clearly seen. In the case of Apollo 15 and 17, similar patterns of low FeO (mostly 10%) and Th (~15 ppm) were seen.

7.1.1.4 *Volcanic activity*

Lunar Prospector data indicated a nonuniform distribution of heat-producing elements in the mantle, which could extend the potential for melting to more recent times than ~3.4–2.2 Byr ago (Hiesinger and Head, 2006). As shown in Fig. 7.3, the lunar basaltic lava covers about 17% of the lunar surface and 1% of the volume of the lunar crust (Head, 1976). Most basalts are located in the nearside basins, forming relatively smooth dark areas, and they appear to be associated with the distribution of KREEP in the Oceanus Procellarum (Haskin *et al.*, 2000; Wieczorek and Phillips, 2000).

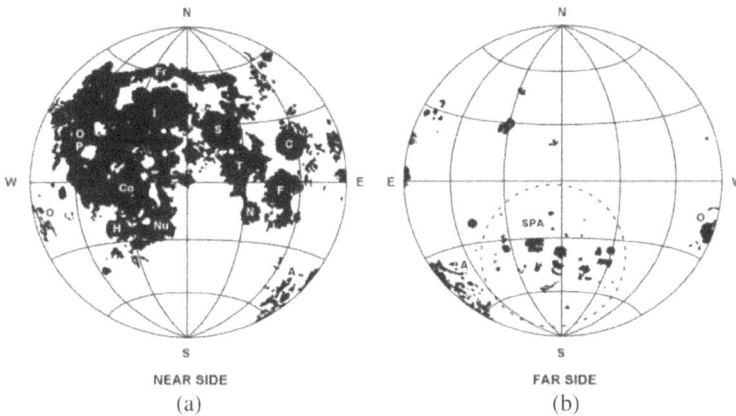

Figure 7.3:   Map of the distribution of mare basalts on the lunar (a) nearside surface and (b) farside surface. The letters indicate volcanic provinces: A, Australe; C, Crisium; Co, Cognitum; F, Fecunditatis; Fr, Frigoris; H, Humorum; I, Imbrium; N, Nectaris; Nu, Nubium; O, Oriental; OP, Oceanus Procellarum; S, Serentitatis; SPA, South Pole-Aitken; T, Tranquillitatis (Head, 1976).

Volcanic features on Moon are identified as lava flows, sinuous rilles, cryptomaria, volcanic centers, domes, sills and shields, cons, lava terrace, and pyroclastic deposits. The thickness of basalts on the lunar nearside is determined to be deep and thicker than 1.5 km (DeHon and Waskom, 1976). The ponds in the SPA basin were interpreted to be single erupted phases having volumes ranging from 35–8745 km^3 — averaging at 860 km^3 (Yingst and Head, 1997).

7.1.1.5 *Water at polar regions*

The M^3 optical spectrometer onboard the Chandrayaan-1 discovered water dependence on the lunar latitude, not only in the polar region but also in the nonpolar regions by observing a water signature with 50–250 ppm concentrations of water molecules and hydroxyl (Pieters *et al.*, 2001). In addition, the recent lunar missions, Lunar Crater Observation and Sensing Satellite (LCROSS) and Lunar Reconnaissance Orbiter (LRO), provided clear evidence of the existence of water in the lunar polar regions (Colaprete *et al.*, 2010; Gladstone *et al.*, 2010; Hayne *et al.*, 2010; Mitrofanov *et al.*, 2010; Paige *et al.*, 2010; Schultz *et al.*, 2010). Using the nuclear method, the abundance of water was determined to be 4.6% and 3.0% for the lunar North and South Poles, respectively. The detection limit of the Lunar Prospector neutron spectrometer (LPNS) was known to be 0.01% for water. The LPNS data confirmed the presence of near-pure water ice deposits 40-cm beneath a layer of dry regolith (Feldman *et al.*, 1998). The impact side of LCROSS inside the Cabeus crater demonstrates the highest hydrogen concentration in the South Pole regions, corresponding to an estimated content of 0.5%–4.0% water ice by weight, depending on the thickness of any overlying dry regolith layer (Mitrofanov *et al.*, 2010). In the case of the KGRS, the hydrogen-signature, which 2223-keV gamma-ray peak in the pole regions was not clearly identified because of interference by a nearby peak, Al (2212 keV) (Hasebe *et al.*, 2009).

7.1.2 *Lunar mantle*

Basalt and volcanic glass compositions combined with experimental petrology can infer the depths of origin and compositions of the mantle source regions (Wieczorek *et al.*, 2006). The major mafic mineralogy (olivine, pyroxene, Fe-Ti oxides) of basalts can be determined from orbit. And the concentrations of FeO and TiO_2 as well as other elements such as Th, K, U, Fe, aluminum (Al), Ca, and magnesium (Mg) can be determined using LPGRS/KGRS data.

REE contents of mare basalt and nonmare crustal materials were key to understanding the basic relationship between the crust and the mantle of Moon (Taylor, 1975). A negative europium (Eu) anomaly can also occur from the crystallization of mafic silicates (McKay *et al.*, 1991). The general variations in REE patterns between mare basalts, feldspathic crustal lithologies, and KREEP-rich materials are consistent with the simple magma ocean scenario (Wieczorek *et al.*, 2006). Mare basalts exhibit distinctive mineralogical, chemical, and isotopic characteristics that reveal major features of their mantle source.

Understanding the trace-element contents of mare source regions is important. The lunar mantle is believed to be extremely dry and volatile-poor (Gibson, 1977). On the basis of the Apollo seismic data, the depth of the upper lunar mantle is found to be roughly 500 km (Khan and Moseggard, 2002). The deep moonquake hypocenters indicated a well-defined source region in the depth range of ~700–1150 km. The transition between the upper and the lower mantle occurred near a depth of 600 km. Below the depth of ~600 km, the mantle seems to be composed of approximately 60 wt% olivine and 40 wt% garnet. The Mg number is found to increase from ~66 in the crust, to ~75 in the upper mantle, and to ~89 at a depth of 600 km (Wieczorek *et al.*, 2006).

7.1.3 *Bulk composition*

Chemical composition data obtained by remote sensing observation, Apollo sample return, and lunar seismic data are key to estimating the bulk composition of Moon. The LPGRS and KGRS measurements provided major elemental maps and natural elemental maps of the Moon (Kobayashi *et al.*, 2010; Prettyman *et al.*, 2006; Yamashita *et al.*, 2010). The general behavior of the elemental abundance maps shows that the nearside mare basalt regions have relatively high Mg abundance and relatively low Al, Ca, and Si abundances. The farside highland regions have relatively low Mg abundance and relatively high Al, Ca, and Si abundances.

The crust and mantle composition of the Moon can be estimated by the bulk composition of the Moon. Th and Al abundances are key to deciphering the bulk composition and origin of the Moon. The bulk composition of Al_2O_3 and abundance of Th are known to be 6 wt% and 125 ppb, respectively (Taylor, 1982).

Recent reanalysis of the Apollo seismic data has revised the crustal thickness at the Apollo 12 and 14 sites down from 60 to 45.5 km (Khan *et al.*, 2000), 39 ± 3 km (Khan and Mosegaard, 2002), and 30 ± 2.5 km (Lognonné *et al.*, 2005). The concentration of Al_2O_3 in the lower crust is known to be less felspathic. The lunar bulk Al_2O_3 abundance in the crust is found to be anywhere between 1.2 and 2.2 wt% for the average thickness of 30–60 km (Wieczorek *et al.*, 2006).

Th concentrations are found to be less than 1 ppm in the upper crust and about 2 or 3 ppm in the lower crust. Assuming the thickness of lower crust to be 25 km, its bulk composition is 1 ppm with an uncertainty of 30 ppb (Wieczorek *et al.*, 2006). A plausible range of bulk Th abundance for the PKT is between 4 and ~7 ppm, corresponding to a 50:50 mixture of KREEP basalt and feldspathic materials.

7.2 Mars

Nuclear planetary science technologies have been applied in Martian exploration to investigate the surface composition of Mars. Science payloads — from the Viking mission to the very recent Mars Science Laboratory (Curiosity) mission — that used nuclear technology have played an important role in the planetary investigation of the elemental features of Martian surface material. This section summarizes the discoveries of Martian features using the science payloads of gamma-ray spectrometers (GRS), neutron spectrometers (NS), and X-ray spectrometers (XRS).

7.2.1 *Mars geological history and features*

With the use of GRS and NS data, it is possible to see very shallow subsurface elemental features up to about several tens of centimeters beneath the surface. This depth is much greater when compared to the measurements using an XRS or optical spectrometer. Mars Odyssey gamma-ray and neutron spectrometer (MOGNS) data have successfully produced major elemental maps of water, Si, Cl, Fe, Th, K, and Ca.

The distribution of K and Th concentrations on Mars was observed by the Mars Observer GRS (MOGRS). The K and Th variations ranged from 2000 to 6000 ppm and 0.2 to 1 ppm, respectively. Th concentration on Mars does not vary as much as that on the Moon (Taylor *et al.*, 2006). The GRS data of K and Th help us to understand the crustal evolution of Mars. On the basis of the K/Th ratio of the MOGRS data, Mars has a higher K/Th ratio than Earth due to its differing temperature environment. Also, low levels of K and Th were found in areas of young volcanic flows, Hadiaca Patera, the area east-southeast of Elysium Mons, Syria, and Solis Planae. This suggests the early crust formation could have been enriched in incompatible elements from later magmas derived from a depleted mantle (Bandfield *et al.*, 2000).

Global maps are useful in understanding Martian geological evolution and current features in comparison with visible or optical remote sensing data. The MOGRS data show that the Si concentrations are varied — mostly from 19% to 22%. In the cases of Tharsis Montes and Olympus Mons, it was found to be about 18%. The uncertainty of Si data ranged

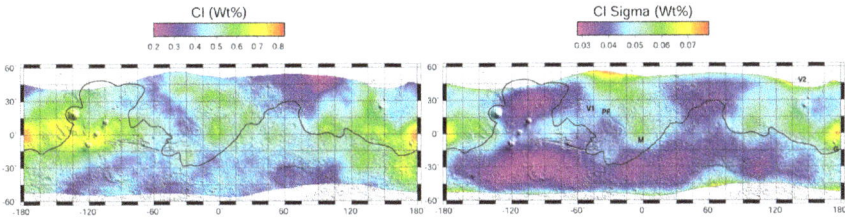

Figure 7.4: Cl elemental map (left) and uncertainty map (right) obtained by Mars Odyssey GRS (Boynton *et al.*, 2007).

from about 0.3% to 0.6%, which is due to the difference in attenuation of the signal from the atmospheric thickness. Then, high-elevation regions tend to have low uncertainties (Boynton *et al.*, 2007).

Global maps of Cl concentration were created using GRS data. The regions with relatively high concentrations were found to be in the Medusae Fossae Formation (see Fig. 7.4). Moderately high amounts of Cl were found in Arabia Terra (Boynton *et al.*, 2007). The MOGRS data show some correlation between Cl and the water signature for the regions of mid-latitude. The highest water concentrations, found over Arabia Terra, in the Medusae fossae Formation, and near Apollinaris Patera, are also associated with high Cl concentrations. The MOGRS data showed that Fe concentrations are distributed from 10% to ~20% with a 0.6%–2.2% uncertainty.

In addition to gamma-ray data, X-ray analysis for surface investigation has a major role in unraveling the bulk composition of Mars, and the geological history of the surface of Mars. From the Viking mission, XRS became a major science payload for surface investigation using a rover or a lander. An active X-ray source has been used for X-ray analysis in Martian experiments because radioisotopes for X-ray fluorescence on Martian surface are simple, light in weight, and high in radiation intensity. Figure 7.5 shows the comparison of elemental ratios for various rocks and soil that were measured by the XRS of Viking and Pathfinder mission samples (Reider *et al.*, 1997). Table 7.1 shows the summary of oxides on Mars and Earth.

7.2.2 *Water on Mars*

The MOGRS data have confirmed the existence of water equivalent hydrogen on the polar and midlatitude regions of Mars (Boynton *et al.*, 2002;

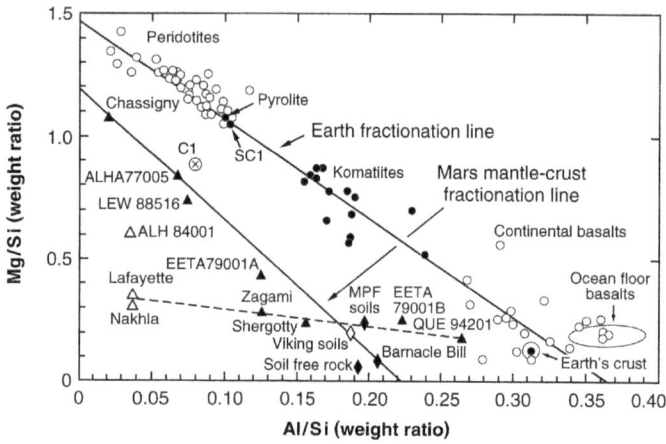

Figure 7.5:   Mg/Si vs. Al/Si diagram of Martian meteorites (filled triangles), mean values of Viking soils (open diamond), and Pathfinder soils (labeled as MPF soils), as well as Barnacle Bill and calculated "soil-free rock" composition (filled diamonds) in comparison with terrestrial samples (Reider *et al.*, 1997).

Table 7.1:   Comparison of oxides from Mars and Earth samples.

Oxide	A-3, Rock "Barnacle Bill"	A-5, Soil	SNCs* Meteorites	Continental Crust Average	Sediments	Oceanic Crust
MgO	3.1	8.6	9.3–31.6	3.1	3.1	7.7
Al_2O_3	12.4	10.1	0.7–12.0	15.2	13.0	15.6
SiO_2	55.0	43.8	38.2–52.7	60.2	50.0	50.7
K_2O*	1.4	0.7	0.022–0.19	2.9	2.0	0.17
CaO	4.6	5.3	0.6–15.8	5.5	8.4	11.4
TiO_2	0.7	0.7	0.1–1.8	0.7	0.7	1.5
MnO*	0.9	0.6	0.44–0.55	0.1	0.1	0.16
FeO	12.7	17.5	17.6–27.1	6.05	5.5	9.9
FeO/MnO	14.1	29.2	37.0–51.5	—	—	—

The header for the table above spans: Mars (wt%) over "Barnacle Bill", A-5 Soil, and SNCs* Meteorites; Earth (wt%) over Continental Crust (Average, Sediments) and Oceanic Crust.

*SNC stands for Shergottite, Nakhlite, Chassigny.
Source: http://nssdc.gsfc.nasa.gov/planetary/marspath/apxs_table1.html.

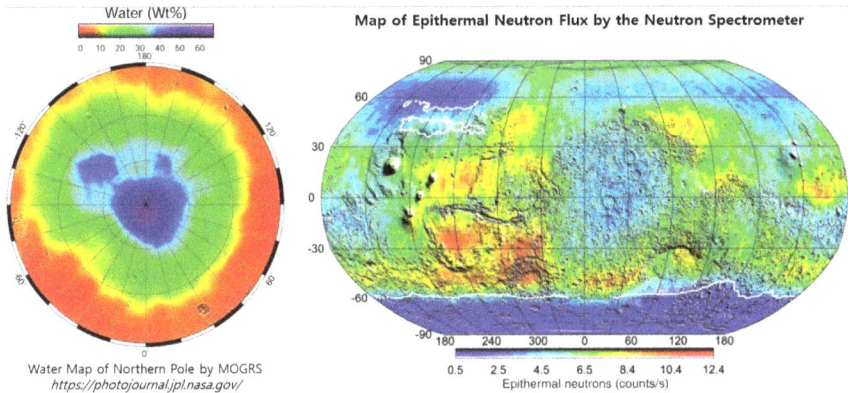

Figure 7.6: Water equivalent maps for polar regions of Mars obtained by Mars Odyssey GRS (Boynton *et al.*, 2002).

2007; Feldman *et al.*, 2002). The highest abundances of water equivalent hydrogen obtained by the MOGRS for the north polar, south polar, and midlatitude regions were more than 90%, 45%, and 7%, respectively (Fig. 7.6) (Boynton *et al.*, 2002; 2007). The water abundance obtained from epithermal neutron data is similar to that obtained from the gamma-ray data, but slightly higher (Feldman *et al.*, 2004b). The existence of water on the polar regions was discovered by the visible camera from Mars Global Surveyor. The results of gamma-ray and neutron spectrometer (GNS) data confirmed it to be hydrogen that is composed of water with the hydrogen gamma-ray peak of 2.223 MeV.

At present, existence of water on the surface of Mars is evident. Prior to which the existence of a large water body on Mars had also, in the past, been confirmed by analyzing the MOGRS data for Th, K, and Fe. The lowland of the northern hemisphere where the elevation is indicated as 0 km (called a shoreline) is found to be enriched with these elements. Three shorelines were suggested by the data on the surface of Mars. The lowland of the northern regions significantly enriched with Th, K, and Fe confirmed the possibility of a paleoocean on Mars (Dohm *et al.*, 2009). This phenomenon possibly occurred because these elements were leached by the water and transported downward to lower land areas. Water equivalent hydrogen in the midlatitude region, where Meridiani Planum is located, has been found to be at least 7% at present (Boynton *et al.*, 2007).

7.2.3 *Atmospheric characteristics*

When a CO_2 cap increases its size in the polar regions, hydrogen gamma-ray flux decreases. Similarly, naturally radioactive element K also decreases with the increase in size of the CO_2 cap. CO_2 depth was, therefore, determined using the signals of H and K in the polar regions. Seasonal variation in CO_2 depth and these nuclides were studied and the variation of CO_2 thickness (g/cm^2) ranged from 0 to about 25 g/cm^2. It was also found that CO_2 increases the flux of thermal neutrons, leading to an increase in gamma rays produced by neutron capture. Therefore, uncertainty in the distribution of H with depth leads to uncertainty in predicting H-based CO_2 thickness (Kelly *et al.*, 2006).

7.3 Mercury

Mercury orbits around the Sun with a sidereal orbital period of 87.65 days. It has a sidereal spin period of 58.65 days, which is two-thirds of the orbital period. This means that two regions of the planet, being 180° apart in longitude, alternately face the Sun at successive perihelia, giving two high-temperature "hot poles." The temperature extremes on the surface are wide and in the range of 100–740 K.

The planet is not symmetrical with respect to its rotating axis. The northern hemisphere is a little more roundish and thus smaller compared to the southern one. Furthermore, like our Moon, no atmospheric gases exist over its surface.

Cratered terrane is found all over the surface of the planet and seems to have resulted from the initial bombardment of large bodies early in the history of the solar system. Observed overlappings of craters in various sizes, similar to those observed over the surface of Earth's Moon, indicate that cratering over the surface took place several times during the history of the planet's evolution.

The magnetic field is intrinsic with a largely dipole form. The strength of one mercurial magnetic dipole moment is estimated to be $(4.9 \pm 0.3) \times 10^{12}$ Tm^3. This would imply an equatorial surface field of 2×10^{-7} T(0.002G) — about one-hundredth of that of Earth (Anderson *et al.*, 2008).

NASA's MESSENGER (MErcury Surface, Space ENvironment, GEochemistry, and Ranging) was launched in 2004 with the objective to observe the surface features of Mercury. It successfully reached near the planet in 2008 and entered orbit around the planet in 2011. Since then, this craft took on the role of mapping Mercury and observing the surface features, in particular, of both the north and the south polar regions to detect some evidence on water ice. At the end of its mission (and two extended missions), after having expanded the last of its fuel supply, MESSENGER finally crashed into Mercury on 30 April 2015 on the farside of the planetary surface. The craft had brought us various important results regarding the nature of the planet. The initial reports about scientific results available from the investigations based on the various observations on Mercury with MESSENGER were reported in the special July 4, 2008, issue of *Science* (and are described as follows).

In this review, the abundant evidence for volcanic activity on the small planet was reported. However, Lawrence *et al.* (2013) reported that the neutron flux was depressed at the north pole of the planet and that high concentrations of hydrogen were confined to the known radar bright locations. These concentrations are consistent with the neutron flux measured at both high and intermediate energies. Thus, the thermal model supports their conclusive results.

The observed results were obtained by both the GRS and XRS on board the MESSENGER. Orbital gamma-ray measurements were done to determine the abundance of the major elements such as Al, Ca, S, Fe, and Na on the surface of the planet (Evans *et al.*, 2012). The Si abundance was determined to normalize those of the other elements referred to above. The Na abundance is estimated to be 2.9 ± 0.1 wt% on the surface of the planet. The other elemental abundances were obtained as $S/Si = 0.092 \pm 0.015$, $Ca/Si = 0.24 \pm 0.05$, and $Fe/Si = 0.077 \pm 0.013$ — consistent with that obtained previously by X-ray measurements.

Although the sampling depths for the two measurements were different, Mercury's regolith is, on average, homogeneous to a depth of tens of centimeters. The elemental abundance obtained from these two measurements is mostly consistent with petrologic models that indicate that Mercury's surface is dominated by Mg-rich silicates.

7.4 Asteroids

A brief review on asteroid research will be given here — especially on the Dawn missions to Ceres and 4 Vesta.

7.4.1 *History*

In 1801, a celestial object orbiting between Mars and Jupiter was discovered by Giuseppe Piazzi (1746–1826). This was thought to be a planet — a hypothesis that satisfied many people including those searching for such a planet. However, this object was soon lost since it failed to pursue its orbit.

On the basis of the least-square method developed by himself, to predict the orbiting motion of the object, Gauss (1777–1855) successfully predicted its position in the celestial sphere. This allowed the small object to be rediscovered fairly soon.

The initial discovery of this object (named Ceres) was thought to verify the Titius–Bode law. But this was the first time that a planet did not agree with this law. The IAU Congress held in 2007 redefined Ceres, as a dwarf planet, together with Pluto (formerly thought to be the ninth planet in our solar system).

Since both the masses and the radii of asteroids are generally so small, asteroids were once thought to have originated from debris produced from the destruction of a planet orbiting between the orbits of Mars and Jupiter. However, it was proven that it was impossible to interpret the origin of asteroids from such a destruction.

A powerfully acceptable hypothesis has recently been proposed by Alexander *et al.* (2012) and Izidoro *et al.* (2014). According to their work, the origin and later evolution of asteroids was causally related to the outward migration of heavy planets, now known as Jupiter and Saturn, in the early history of the evolution of the planetary system. Due to the migratory motion of these two planets, the enormous amount of masses that prevailed in the current space between the orbits of Mars and Jupiter were taken away by the outward migration of Jupiter and Saturn. In relation to this process, the original mass to be consumed to build up Mars was lost so effectively that the mass of Mars became so tiny as compared to that of

Earth and Jupiter. Throughout these processes, the masses of both Jupiter and Saturn increased so much compared to those for inner planets such as Earth and the other terrestrial planets.

7.4.1.1 *Classification of asteroids*

There are millions of asteroids, many thought to be the fragmented bodies of planetesimals. The large majority of known asteroids orbit in the asteroid belt between the orbits of Mars and Jupiter called the main belt. Individual asteroids are classified by their characteristic spectra, with the majority falling into three main groups: C-type, M-type, and S-type, named after and are generally identified with carbon-rich, metallic, and silicate (stony) compositions, respectively. About 80% of asteroids belong to either the C-type or S-type groups. The former is mostly classified as carbonaceous chondritic substance, whereas the latter is mainly built up with Fe- and Mg-rich silicates. C-type asteroids are the most common, are carbonaceous, and consist of clay and silicate rocks. S-type asteroids show spectra with 1 and 2 μm absorption features indicative of olivine and pyroxene, and slopes indicative of metal. They are made up primarily of stony materials and Fe-Ni and inhabit the inner asteroid belt. M-type asteroids show flat and featureless spectra with intermediate albedos. They are made up mostly of Ni-Fe and are found in the middle region of the asteroid belt. The dominant composition of main belt asteroids change with heliocentric distance, which suggests that they have remained near their original formation regions since the origin of the solar system.

Near-Earth asteroids (NEAs) have semi-major axes ranging from <1 to 1.5 AU. The majority of the present NEAs are believed to have been perturbed from the main belt into planet-crossing orbits.

7.4.2 *Recent exploration: Dawn spacecraft*

7.4.2.1 *Dawn mission and gamma-ray and neutron detector*

A brief review of the Dawn spacecraft's CRES and Vesta asteroid missions is given here. The mission aimed to observe two massive objects, Ceres and Vesta. Though Ceres is now classified as a dwarf planet, the observations of these objects might still give us some insight into bodies with rocky and icy interiors. NASA's Dawn spacecraft launched in September 2007 to study the two protoplanets, Vesta and Ceres, and entered orbit around Vesta on 16 July 2011 for a 1-year exploration before leaving its orbit of Vesta on 5 September 2012 for its final destination, Ceres. It is the first NASA exploratory mission to use ion propulsion. Dawn carries visual and infrared (VIR) framing cameras (FCs), a VIR spectrometer, and a gamma-ray and neutron detector (GRaND). One of the largest objects in the asteroid belt is 4 Vesta, with a mean diameter of 525 km. Vesta is the second largest body in the asteroid belt. The perihelion and aphelion of Vesta's orbit are 2.32 and 2.57 AU, respectively. Vesta is thought to be a remaining rocky protoplanet with a layered differentiated interior structure. A large number of fragments were ejected from Vesta by collisions that left two enormous craters, especially in the southern hemisphere of Vesta. The most prominent of these surface features are two enormous craters, the 500-km-wide Rheasilvia crater, centered near the south pole, and the 400-km-wide Veneneia crater. The Rheasilvia crater is younger and overlies the Veneneia crater.

Dawn carried the GRaND to measure and map the elemental composition of Vesta (Prettyman *et al.*, 2011; 2015a). Elemental data were accumulated and the elemental maps were created as shown in Fig. 7.4 (Prettyman *et al.*, 2015b). Mapping of neutrons in the epithermal energy range (approximately 0.5 eV to 0.5 MeV) revealed elevated concentrations of H within Vesta's "dark" (low albedo) hemisphere (Prettyman *et al.*, 2012).

7.4.2.2 *Relation between 4 Vesta and howardite-eucrite-diogenite meteorites*

Some of fragments ejected from Vesta have fallen to Earth as howardite-eucrite-diogenite (HED) meteorites, which have been a rich source of information about Vesta. The HED–Vesta relation is supported by a close spectroscopic approach between Vesta and the HEDs (Binzel *et al.*, 1997; Li *et al.*, 2010; McCord *et al.*, 1970). Studies of the HEDs show that their parent body underwent igneous differentiation to form a metallic core, ultramafic mantle, and basaltic crust (McSween *et al.*, 2013).

Spectra of asteroid 4 Vesta are characterized by strong absorption bands of FeO-bearing pyroxenes that are indistinguishable from spectra of HED meteorites. The identification of small Vesta-like asteroids has led to the consensus that Vesta is the HED parent asteroid (Binzel and Xu, 1993).

One of the main objectives of the Dawn mission was to test the Vesta-HED connection hypothesis and to provide geologic context for the HEDs (McSween *et al.*, 2013). Dawn observed it for over a year in proximity to Vesta (Russell *et al.*, 2012), enabling detailed studies of Vesta's composition, geology, geomorphology, and internal structure. Global elemental concentrations (Fe/Si and Fe/O weight fraction ratios) determined by the GRaND are consistent with the HED whole-rock compositions, further strengthening the link between Vesta and the HEDs, which was confirmed by the use of fast neutrons (>0.5 MeV). The GRaND observations support the argument of Vesta being the source of HEDs, giving us new insight into a differentiated asteroid (McSween *et al.*, 2013).

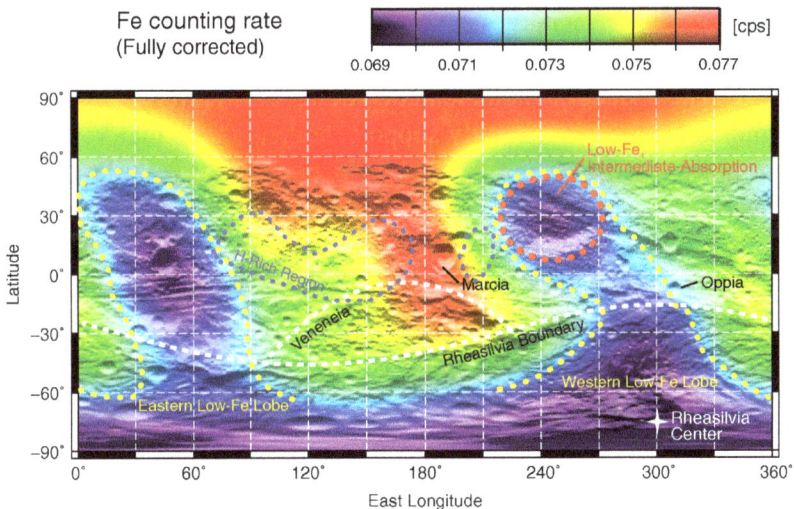

Figure 7.7: Global distribution of Fe counting rates on the surface of Vesta. Eastern and western low-Fe lobes are delineated with yellow dots. H-rich regions are shown with blue dots. The low-Fe, intermediate-absorption region, indicative of cumulate-eucritic materials, is presented with red broken circles (Yamashita *et al.*, 2013).

7.4.2.3 *GRaND observation*

Iron can be used along with other mineralogic and chemical information to classify geologic units of surface material. Basaltic eucrites have the highest Fe abundance, whereas lower crustal and upper mantle materials (cumulate eucrites and diogenites) have the lowest. Howardites have intermediate abundance with a range of about 12–17 wt% Fe. The abundance of Fe in olivine-rich rocks depends on the concentration of the forsterite component in olivine (Prettyman *et al.*, 2013; Yamashita *et al.*, 2013). Global Fe/O and Fe/Si ratios are consistent with HED compositions. Neutron measurements confirmed that a thick, diogenitic lower crust is exposed in the Rheasilvia basin, which is consistent with global magmatic differentiation. Vesta's regolith is found to contain substantial amounts of hydrogen. The highest hydrogen concentrations coincide with older, low-albedo regions near the equator where water ice is unstable, whereas the young, Rheasilvia basin contains the lowest concentrations.

The concentrations of K and Th within Vesta's global regolith measured by GRaND are consistent with eucrite-rich howardites, and are distinct from most achondrites, all chondrites, and Mars meteorites. The K/Th ratio of Vesta (900+/−400) is similar to the average ratio for howardites (~1200). These data of radioelements along with major element ratios determined by GRaND strongly support the hypothesis that Vesta is the parent body of HEDs. The depletion of moderately volatile elements implied by the measured K/Th ratio is consistent with an early accretion of Vesta from a hot and incompletely condensed solar nebula, and/or subsequent removal of volatiles by energetic collisions or degassing of magmas (Prettyman *et al.*, 2015b).

The Dawn spacecraft entered low-altitude mapping orbit (LAMO) around Ceres — a circular, polar orbit with a mean altitude of 0.8 Ceres body radii. GRaND accumulated 2 weeks of mapping data, with regional-scale resolution and full global coverage. Preliminary results indicate that Ceres' regolith has an H-rich, carbonaceous chondrite-like composition, with water ice likely present near the surface at high latitudes. More detailed results of the analysis and implications for Ceres' formation and evolution will be reported in the near future.

7.5 Future Missions

7.5.1 *Future mission for Korea*

7.5.1.1 *Introduction to the science payloads for the Korean*
Pathfinder Lunar Orbiter

Korea is planning its own lunar mission in two stages. The first one is called the Korean Pathfinder Lunar Orbiter (KPLO) that is planned to be developed by 2018 and the second lunar program is currently under investigation. The second spacecraft is going to be launched with a Korean rocket and it is planned to have a lander with a rover. For these missions, lunar geological investigation and resource exploration will be one of the major important tasks for current and future interests in conjunction with prospective permanent lunar activities by humans. For the KPLO, five payloads were selected to be developed in Korea and two payloads are supposed to be developed by NASA. The KPLO will carry a high-definition camera, a suite of instruments including a pantoscopic polarizing camera, a GRS, and a device to measure the Moon's magnetic field and well as the 'space Internet' — a delay tolerant network (DTN).

The orbit of the spacecraft will be 100 km above the lunar surface. The nominal mission period is 1 year, including a 1 month calibration period. The purpose of the KPLO is to investigate geographical features of the Moon, the lunar environment, and the lunar resources from the near surface. The high-definition camera, LUnar Terrane Imager (LUTI), can make an image of the surface as small as 5 m of the lunar surface. The pantoscopic polarizing camera (Polcam) will observe polarizing light from the lunar nearside and investigate the sizes and types of particles encountered on the lunar farside, as well as high-energy particles from the Sun. The pantoscopic images of the entire lunar surfaces will assist in finding a suitable landing site for the future Korean lunar mission. The magnetometer of the KPLO will be used to measure the lunar magnetic field, which has been gradually disappearing since the early stage of the Moon. The KPLO magnetometer (KMAG) is targeted to provide a high-precision 3D map of the lunar magnetic field. This could give new information on the formation of the Moon and its early evolutionary process. The KGRS is to determine the composition and distribution of elements on the lunar surface by examining gamma rays emitted from the lunar surface. Various elemental maps of the lunar surface and resources will be created with the GRS.

The space Internet, also known as the interplanetary Internet, is a proposed network intended to connect stations on Earth with others in orbit around and on the surface of other planets and moons in the solar system. DTN characteristics and protocols will be tested during the lunar exploration of the KPLO. The proposed DTN test architecture is based on the Lander and Rover located on earth. In the second phase, the Lander and Rover will be landed on the moon. The DTN architecture and protocol stacks for the second phase are also proposed.

7.5.1.2 *GRS for the KPLO*

GRS is a widely used technique to map and understand the surface composition and evolution of planetary surfaces. It has been used on several planetary missions

As a science payload related to the nuclear planetology, a lanthanum(III) bromide ($LaBr_3$) system onboard KPLO will investigate both lunar geology and lunar resource. This GRS system will allow new investigations on the elemental mapping of REEs and volatile elements as well as other precious metal in the lunar polar regions by obtaining X-ray and wide range of gamma ray spectra using the following:

MARS: Mars Observer (HPGe) [1993, NASA], Mars Odyssey (HPGe) [2001–2006, NASA]

MOON: Apollo 15 and 16 (NaI) [1971/1972, NASA], Lunar Prospector (BGO) [1998, NASA]

            SELENE-1 (Kaguya) (HPGe) [2007–2009, JAXA], Chang'E 1 (CsI) [2007–2009, CNSA] and Chang'E-2 ($LaBr_3$) [2010–2011, CNSA]

MERCURY: MESSENGER (Mercury) (HPGe) [2011, NASA], BepiColombo ($LaBr_3$) [2024, ESA]

ASTEROIDS: Dawn (Ceres, Vesta) (BGO) [2011–2015, NASA], NEAR (Eros) (NaI) [1996, NASA]

The response of the $LaBr_3$ GRS instrument of the KGRS (Fig. 7.8) and its concept of the operation of a GRS is shown in Fig. 7.9.

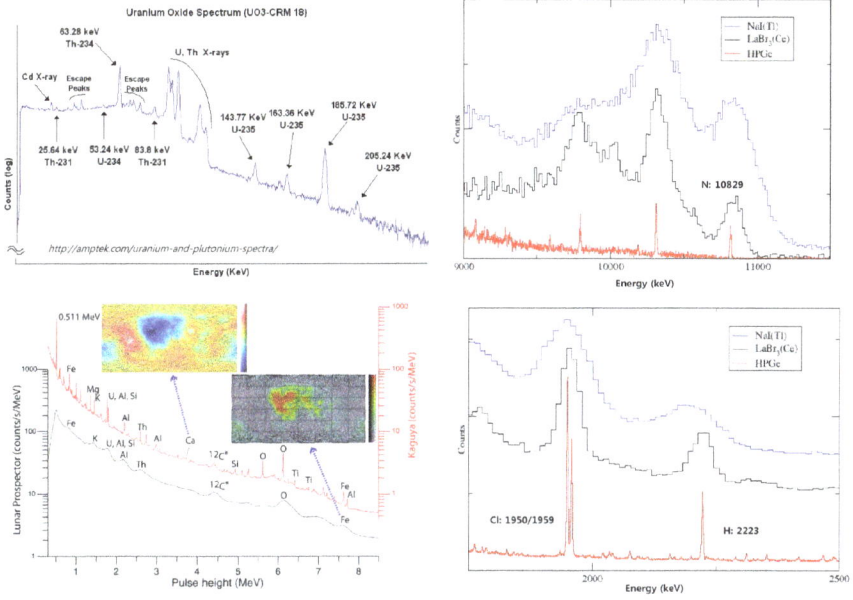

Figure 7.8: Examples of **GRS** spectra of a LaBr$_3$ detector (Seabury *et al.*, 2006; Zhu *et al.*, 2015).

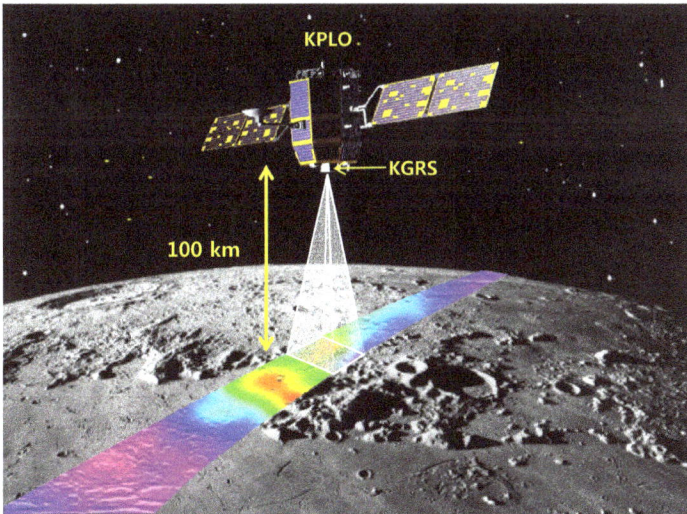

Figure 7.9: Concept of a **GRS** on the Moon.

7.5.1.3 *LaBr GRS detector system for the KPLO*

7.5.1.3.1 Major scientific and technical objectives

The LaBr$_3$ GRS system is a recently developed gamma-ray scintilla-tion-based detector that can be used as a replacement for high-purity germanium (HPGe) GRS sensors with the advantage of being able to operate at a wide range of temperatures with remarkable energy resolution. LaBr$_3$ also has a high photoelectron yield, fast scintillation response, good linearity, and thermal stability. With these major advan-tages, the LaBr$_3$ GRS system will allow us to investigate fundamental scientific topics and assess important research questions on lunar geol-ogy and resource exploration. The latter has become of importance, in particular, as recent results of the LCROSS/LRO mission confirmed lunar resources, including water, volatiles, and precious metals (e.g., Hg 1 wt% in lunar soils), which can be utilized for future human settlement on Moon. The proposed LaBr$_3$ GRS system is focused on (1) mapping lunar resources located in the polar regions, (2) elemental mapping on a global scale, and (3) mapping of the radiation environ-ment in the X-ray to gamma-ray energy range corresponding to tens of keV to 10 MeV. The number of energy peaks from the GRS spectra can be used to identify elemental abundance on the lunar surface when the interfering background gamma rays are reduced by an anticoincidence counting system (Fig. 7.10).

The KGRS aims to take a new approach to collecting low-energy gamma-ray signals; to detect elements by either X-ray fluorescence or natural radioactive decay in the low as well as higher energy regions

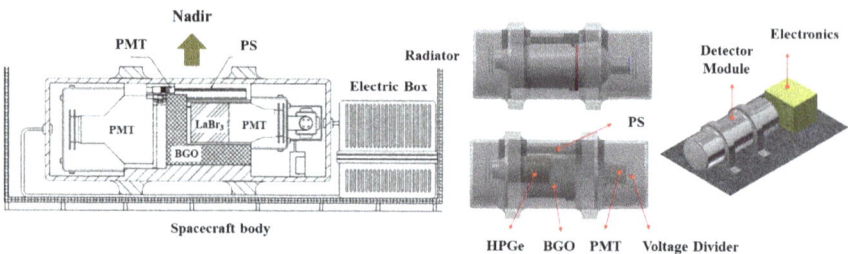

Figure 7.10:   Preliminary cross-sectional view of a LaBr$_3$ GRS system and components for a future lunar mission.

of up to 10 MeV. Scientific objectives include but are not limited to the following:

Lunar resources: (1) water and volatile measurements in polar regions, (2) REEs and precious metals in polar regions (X-ray regions), (3) energy resources such as U and He-3, and (4) major elemental distributions for prospective *in situ* utilizations.

Lunar geology: (1) Mg distribution and lunar impact history and evolution, (2) mineralogical distribution and surface evolution through elemental mapping, and (3) lunar surface activities by monitoring of X-ray release by ^{222}Rn and ^{210}Po.

Lunar environment: (1) mapping of the global radiation environment from tens of keV to 10 MeV (X-ray and gamma-ray regions) and (2) high-energy cosmic ray flux using the plastic scintillator.

7.5.1.3.2 Instrument description and specification

The LaBr$_3$ GRS system is a compact low-weight instrument for the chemical analysis of lunar surface materials within a gamma-ray energy range from 10 keV to 10 MeV (Fig. 7.10). The main LaBr$_3$ detector is surrounded by an anticoincidence counting module of bismuth germinate or a plastic scintillator (BGO/PS) to reduce (1) the low-energy Compton gamma-ray background from the spacecraft and housing materials and (2) the high-energy particle background from cosmic ray particles (Fig. 7.10, Table 7.2). To minimize prompt gamma-ray background, Ti is used as housing material. Monte Carlo N-Particle eXtended (MCNPX) numerical simulation will be used for the optimization of the detector design.

The LaBr$_3$ GRS system is a compact low-weight instrument for the chemical analysis of lunar surface materials within a gamma-ray energy range from 10 keV to 10 MeV. A higher number of channels and lower gamma-ray energy measurement will provide more distinct results when compared to earlier instruments, such as the Chang'E-2 GRS.

7.5.1.3.3 Instrument development

Both national and international teams are working together on technical developments and scientific applications for KPLO. The mechanical

Table 7.2:   GRS system specifications for a future lunar mission.

Component	Main det. (LABR$_3$)	Shielding det. (BGO)	Shielding det. (PS)
Type	Scintillator	Scintillator	Scintillator
High voltage (HV)	1000	800	200
No. of channels	8192 (or TBD)	1024	512
Energy resolution	<3.00% @662 keV (FWHM)		
Energy range	0–10 MeV, 1.22 keV/channel		
Size	3″ × 3″ (or 4″ × 4″ TBD)	Cylindrical/horseshoe	0.5-cm thickness
Housing for detector	Material-Ti (TBD)		
Power	10 W (regulated-DC + 28 V)		
Total mass and size	5 ± 0.5 kg, 20 × 20 × 30 cm^3 (TBD)		
Attachment/direction	Bottom of SC/nadir		
Operation temperature	20 ± 10°C (stability of LaBr$_3$ sensor at –30 to 70°C)		
Data accumulated/volume	20 s/1.5 Gbits/day (compressed when required)		
Data rate	35.76 kbps (anti-coin) 8192 channel. TC/TM interface: RS-422		

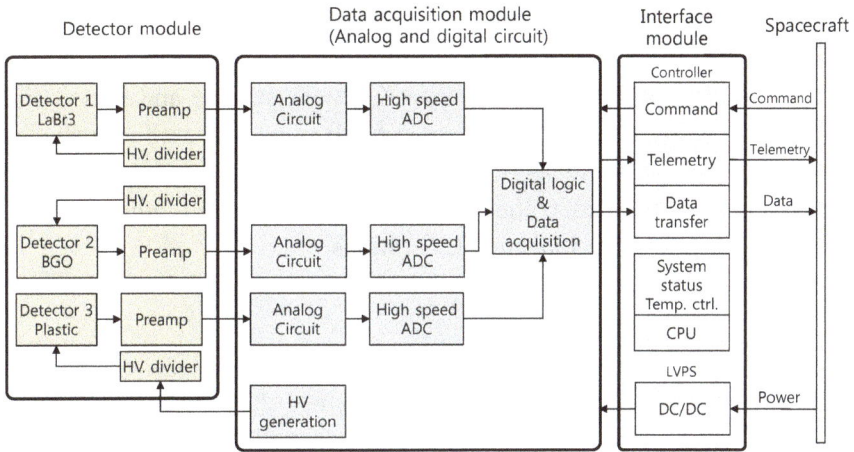

Figure 7.11: Preliminary scheme of the electronics layout.

and thermal designs are to be developed by aerospace engineers. In the course of development, different sets of GRS units (PM, EQM, FM) will be designed, constructed, and evaluated. The GRS sensor, including the anticoincidence counting system, is being made at the Korea Institute of Geoscience and Mineral Resources (KIGAM). All electronics will be made by a Korean company that designs and develops radiation and nuclear instruments (Fig. 7.11). The performance of each GRS model will be tested with calibration sources in a lunar analog environment.

The LaBr$_3$ GRS system shall be operated at all times unless other instrument operations are prioritized. The accumulation time for a GRS spectrum is approximately 20 s. The daily data volume is expected to be about 1.5 Gbits/day with one channel. The GRS data should be provided along with positional data at all times. Topographic data available from the Planetary Data System (PDS) are required for corrections and elemental mapping.

The energy calibration of the LaBr$_3$ detector's gamma-ray spectrum is performed using a mixed gamma-ray source. Its energy resolution is to be checked to ensure the reliability of its GRS measurements. In flight, the intrinsic gamma-ray lines from the LaBr$_3$ crystal are used to check and validate the energy peak identification and any shift due to changes within the cosmic environment or temperature. For background subtraction, it is of utmost importance to take a background spectrum before the nominal mission.

The LaBr$_3$ GRS system will face the lunar surface and must point nadir in mapping mode. It is desirable for it to be mounted in a shaded zone of the spacecraft when the spacecraft is in orbit. A thermal model analysis that also considers the locations' temperature and radiation backgrounds, is required to find suitable locations.

7.5.1.4 *Summary*

The GRS science payload has been used since the Apollo 15/16 lunar missions in the early 1970s. Since then — and well into the course of the recent Chang'E-2 mission — the major direction of focus for GRS instrument development was lunar geology and elemental mapping, in order to understand the evolution of the Moon. However, recent developments target not only the investigation of lunar geologic history. Today, both lunar geologic investigation and *in-situ* analysis and future utilization of desirable resources are important for future planetary research applications. Researchers hope to expand the energy range of future GRSs to measure not only gamma rays but also (hard) X-ray energy regions that are associated with REEs (above 20 keV). In addition, the utilization of volatiles including water in the polar region is important; therefore, expanding the measurable energy range and the number of measureable X-ray and gamma-ray regions should be considered. No matter what type of GRS instrument is chosen, fulfilling the above-mentioned requirements is of importance in order to contribute to an innovative and sustainable exploration program. In the case of KPLO's LaBr$_3$ GRS detector system, measurements of volatile elements and nuclides that emit X-rays should be considered. To tackle these challenging tasks, more advanced techniques in both background reduction and signal processing modules need to be developed. Unlike previous GRS instruments used in lunar surface exploration, KPLO's KGRS focuses much more on finding lunar resources. The results of this effort could offer a guide to humans regarding the obtaining of lunar resources on the Moon; thereby assisting in enabling the utilization of lunar resources for future human activities on the Moon.

7.5.2 *Martian phobos/deimos mission*

The Martian moons, Phobos and Deimos, have never been explored before, so the chemical compositions of these moons are still unknown. Moreover, there is also a lack of high-resolution optical observations, especially in two key wavelengths: 0.65 and 2.8 μm (Basilevsky *et al.*, 2014; Schmedemann *et al.*, 2014; Thomas *et al.*, 2010; Witasse *et al.*, 2014). The origin of these moons is controversial (Citron *et al.*, 2015; Craddock, 2011; Rosenblatt and Charnoz, 2012). At present, the Japanese mission, Mars Moon eXploration (MMX), currently at the planning stage, plans to make close-up remote sensing and *in-situ* observations of both moons and return samples from Phobos to Earth. The major scientific objectives of MMX are to determine whether the origin of these moons is of the captured asteroid or giant impact type. Remote sensing, including nuclear spectroscopy, will characterize the Phobos' surface in advance before landing — providing useful information regarding the landing sites for sample collection. The multi-instrument remote sensing coverage of the moons by the MMX also contributes to enlarging our understanding of formation processes at the outer edge of the rocky region in our solar system.

7.5.2.1 *Origin of martian moons, Phobos and Deimos*

The origin of Phobos and Deimos is itself a nice question to be answered. At present, there are two leading ideas for the origin of the two moons: (1) that Phobos and Deimos are asteroids 'captured' by Mars or (2) that the two moons were formed as a result of a giant impact Mars suffered with a protoplanet. The former requires the orbital energy of the captured asteroids to be dissipated within the Mars gravitational sphere. The latter assumes that a moon was accreted from a debris disk around Mars formed due to the planet sustaining a sizeable impact. At present, there is no consensus regarding the result of the contest between the two ideas, and neither can satisfy all the constraints derived from existing data.

Characteristic properties of the moons related to their origin are as follows: (1) their orbits are mostly circular (eccentricity $\varepsilon = 0.01511$ for Phobos, and 0.00024 for Deimos) and mostly on Mars' equatorial plane (inclination $I = 1.08°$ and $1.79°$, respectively); (2) Phobos is inside and Deimos is outside the corotation radius; and (3) their average mass densities are lower than the typical values for rocks (1.85 and 1.48 g/cm^3, respectively; compared to rocks, which range from 2.5–3 g/cm^3). The orbits of the moons are well known, with Phobos located inside of the corotation radius ($a_c = 5.9 \times R_m$, where R_m is the radius of Mars) drifting inward toward Mars, and with Deimos outside of the corotation radius moving outward from Mars. These facts imply either the presence of ice or high internal porosity in the bodies.

From limited near-IR spectroscopic observations, both Phobos and Deimos have low reflectance and monotonic spectrum shapes, resembling D-type asteroids. Most importantly, there is a lack of observations at the two key wavelengths, 0.65 and 2.8 μm, as mentioned before.

Though the amount of remote sensing data for Phobos is not small, they are not enough to definitively determine which of the two ideas for its origin is the right one. The MMX is studying material from Phobos obtained by an earlier a sample return mission. It is likely that remote sensing data alone would not lead us to a definitive conclusion on Phobos' origin. Return samples from Phobos, which include samples of Phobos' original building blocks, are a good way to obtain an answer to the question.

Close-up observations of Phobos and Deimos would enable us to give a strong constraint to the idea of their origin. If the origin of Phobos is assumed to be a captured primordial asteroid, a detailed analysis of the returned samples allows us to study how the primordial materials, namely water and organic compounds, were brought into the inner part of the solar system from the outer part across the snow line. And if the moons were formed by a giant impact with Mars, their composition is expected to lack volatile elements such as water. The measurement of globally averaged elemental abundance ratios of Ca/Fe and Si/Fe will be a good measure of the constraint of their origin. And sulfur S will also give a good indicator of their origin.

7.5.2.2 *Gamma-ray and neutron spectrometer*

The MMX payload will carry a gamma-ray and neutron spectrometer (GNS) that measures and globally maps the surface elemental composition of the two moons (Hasebe *et al.*, 2016).

The GNS combines the distinct features of being lightweight, having low power requirements, and possessing excellent energy resolution. It allows us to assess the composition maps of elements such as H, O, Mg, Al, Si, S, K, Ca, Ti, Fe, Th, and U, depending on their concentrations in the Martian moons. The high concentration of volatile elements such as H and S in Martian moons would show that they are solar system primordial bodies, whereas high values of Ca/F and Si/Fe ratios and very low water would suggest that they originated from a giant impact. The GNS is expected to play an important role in the MMX mission to obtain the elemental composition of the moons' surfaces by remote observation before landing.

7.5.2.3 *Expected results*

The scientific objectives of the gamma-ray and neutron spectrometer (GNS) are to make contributions toward revealing the origin of the Martian moons. To determine whether Phobos originated from a captured asteroid or a giant impact, the materials that constitute the moons need to be characterized. Measurement of globally averaged elemental abundance of O, Mg, Al, Si, Ca, Ti, Fe, K, Th, and U, and the ratios of Ca/Fe and Si/Fe will characterize the materials of Phobos and Deimos and lead us to give a constraint to their origin (Hasebe *et al.*, 2016; Yoshida *et al.*, 2016; Naito *et al.*, 2016; see Fig. 7.12).

We assume two different types of elemental compositions as possible material characterizing the Martian moons. One is the chondritic composition (C-type: C, CM, CO, and CI meteorite; Jarosewich, 1990), which would support the capture origin argument; and the other is a Martian composition (Shergotty and Nakhla meteorites; NASA, 2016), which would support the giant impact origin argument.

Si/Fe and Ca/Fe ratios are useful indicators for differentiating chondritic from Martian composition. CI chondrites enriched with volatile elements such as H and S, and depleted of Ca and Si are taken in the numerical simulations as the most primitive chondrites. Shergotty meteorites, which are the majority of Martian meteorites, are depleted of volatile elements because of impact heat and have Si and Ca in abundance.

From the results of the numerical calculation of Si/Fe and Ca/Fe ratios, three elements, Si, Ca, and Fe, show a clear separation between the

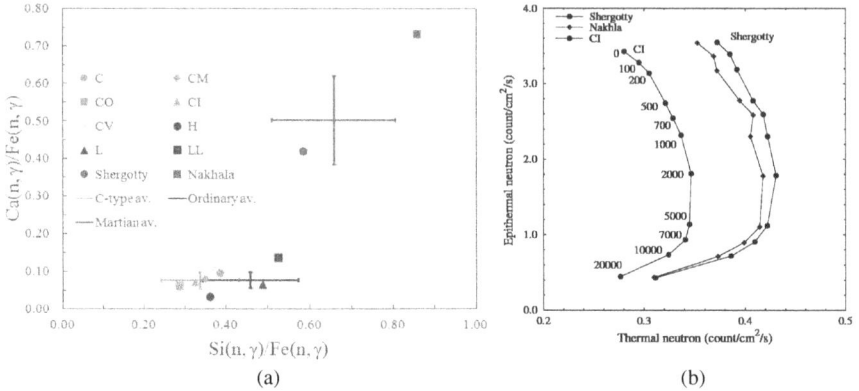

Figure 7.12:   Ca/Fe vs. Si/Fe ratio for 1 and 2 weeks' observations by GRS at an altitude of 20 km from Phobos' surface (left). The relation of neutron fluxes between thermal (LiG) and epithermal neutrons (boron-loaded plastic, BLP) counting rates (right).

two ideas of origin for the Martian moons. In this calculation, the MMX is assumed to move around at an altitude of 20 km. Figure 7.12 shows the emission flux of gamma rays from Phobos (left) and the energy spectrum of neutrons emitted from the surface (right).

7.5.3 *NEA missions*

7.5.3.1 *Motivation*

Space activities using nanosats/microsats (nanosatellites and micros-atellites) are rapidly expanding around the world. With recent progress in the miniaturization and increased capabilities of advanced electronic, material, and information technology, microsats and nanosats have higher potentials to perform attractive missions. There has been increasing demand for small satellites so as to fulfill users' requirements for "low cost, quick delivery, and high performance." These demands increase for the systems combining satellites, rockets, ground stations, data services, and training. Microsatellites have a huge potential for innovation in technology demonstrations, and enable the utilization of space systems by realizing very low-cost and short-time delivery, despite some limits in product lifespans and resolution accuracy. In this section, we describe nuclear spectrometer on the scale of 50- to 100-kg-class microsatellites for the purpose of developing early proof-of-concepts for future deep space exploration through the study of NEAs by multiple Earth swing-bys.

With the use of a miniature deep space probe — a 50–100 kg class microsat, rendezvous missions for NEAs (near-Earth asteroids) would be very interesting next steps for deep space exploration (Funase *et al.*, 2014; Koizumi *et al.*, 2012; 2014; Ozaki *et al.*, 2014). Such microsats are able to have a small and lightweight engines (e.g. a Xe-ion engine) and can approach small bodies moving near the orbit of Earth. Moreover, they can offer a very quick turnaround and an inexpensive means of exploring well-focused, small-scale science objectives, or provide an early proof of concept prior to the development of large-scale instrumentation in a fully complementary manner to expensive, large-scale planetary science missions.

This not only yields early and quick scientific data but also provides opportunities for students pursuing master's and doctoral courses, young scientists, and engineers, to gain "real-life" experience with satellite and payload engineering (an invaluable experience for later large-scale missions). As an education program in a graduate course, a doctoral student can initiate a program researching, proposing and building an instrument, for the retrieval of orbital data for analysis and presentation for a thesis, within a normal period of postgraduate study. In fact, recently, there have been issues raised, centering on the need to promote space activities by universities for the above reasons. Driven by their own vision and efforts, university teams could launch their own satellite into space to reach a new horizon for space research.

Nuclear spectroscopy is widely applicable to the space exploration of near-Earth objects (NEOs). However, in this book, we tentatively take up the exploration of an M-type asteroid, among various asteroids. The exploration, by proxy, is to survey the surface and interior structure of terrestrial planets and moons. While we cannot visit the core any other way, it has become relatively easy to access such a small NEO with the rise of technology like nanosats and microsats, which facilitate low-cost rendezvous missions. The missions would characterize an asteroid's geology, shape, and elemental and mineralogical composition (Hasebe *et al.*, 2014; 2015).

7.5.3.2 *Scientific background*

7.5.3.2.1 Core formation of terrestrial planets

Terrestrial planets are formed through an accretion process. The accretion makes tiny solid particles grow into planets. In the early phase, these particles are able to stick together through electrostatic forces, not through gravitational forces, because they are too small. But as they grow in mass, their gravitational forces will accelerate their growth, forming planetesimals. And these planetesimals keep growing and rapidly joining together until they form a mass of a decent planetary size. They accrete as chondritic mixtures of metal and sulfide grains, minerals of relatively low-melting temperature, and abundant high-melting Mg and Fe silicates. The internal heating of the new planet causes the mineral mixture to melt partially. Molten metals and sulfides move to the centers, forming cores. Silicates that melt at low temperatures produce a low-density magma, which tends to erupt from the planetary surface. From an early stage, terrestrial planetary bodies already form a core with an upper mantle mostly made up of silicon oxide or silicate and a mixture of Fe and Mg. At high temperatures, which may be due to the radioactive decay of short-lived isotopes, a metallic melt containing Fe forms between grains of silicate crystals, because it has a lower melting point. The separation from the core could have occurred very quickly, probably in less than a few million years.

7.5.3.2.2 Earth's core

The average density of Earth is 5.5 g/cm^3, whereas the average surface density is only around 3.0 g/cm^3. Earth's core is thought to be composed of heavy materials. The core can be divided into two parts: a "solid" inner core (12.6–13.0 g/cm^3) with a radius of ~1220 km and a "liquid" outer core (9.9–12.2 g/cm^3) extending beyond a radius of ~3400 km. The core is thus believed to be composed largely of Fe along with Ni, and other light elements.

Dynamo theory suggests that convection in the outer core gives rise to the Earth's magnetic field. Moreover, recent speculation suggests that the innermost core is enriched with Au, Pt, and other elements (Wootton, 2006), because the inner core is denser than pure Fe/Ni, even under high pressures. Precious metals and other heavy elements are expected to be more abundant in Earth's inner core than in its crust and mantle.

Recent investigations of Earth's structure have been closely related in their study of both the core's chemical composition and the convective motion of superplumes inside the mantle, since the existence of light elements such as O, Si, and other refractory lithophile elements has been required to understand the physical properties of the outer core (Duffy. 2011; Huang *et al.*, 2011; Ishida *et al.*, 1999; Maruyama *et al.*, 1994). Recent seismic data provide constraints on density and sound velocities throughout the core. These elements may have played an important role in the superplume motion deep inside the mantle transferred from the outer core. Light elements such as H, O, and Si, which are being accumulated in the outer core, may efficiently drive up superplume accumulated on the mantle/core boundary (MCB) and transferred from the MCB to Earth's crust.

Earth's liquid outer core may consist of mainly liquid Fe/Ni alloyed with about 8 wt% of light elements such as S, O, Si, C, and H (Birch, 1964). But the precise identities of those in Earth's core are still unclear. New progress in "Plume tectonics" proposed to explain not only the superficial layer but also the dynamics in the whole system of the earth (Buffett, 2000; Maruyama, 1994).

Better knowledge of the core's elements — not only the siderophile elements but also light elements such as S, Si, O, C and H — will shed light on heat flow in Earth's deep interior, the origin and growth of the core's solid inner regions, and the generation and evolution of the geomagnetosphere (Duffy, 2011). An understanding of the detailed structure and internal motion inside the mantle and the outer core is required to take into account the distribution of light elements in the outer core.

The elemental abundance of M-type asteroids will shed light on our understanding of the earth's core because Earth's core is inaccessible; thus preventing direct research. The real purpose of space exploration is to know the earth better.

7.5.3.2.3 Iron meteorites

Iron meteorites are nearly 100% metallic, although many contain the FeS mineral troilite. Stony-iron meteorites consist of olivine grains embedded in an Fe/Ni metal matrix. Naturally occurring terrestrial rocks do not contain iron metal or Fe/Ni metal. Earth rocks only contain Fe and Ni in their oxidized (nonmetallic) forms. Early in Earth's history, Fe/Ni metal was thought to be metal that had sunk to form Earth's core.

Iron meteorites are thought to have been originally formed as cores of a parent body. Therefore, the observation of the parent body of an iron meteorite by remote sensing is significantly equivalent to a direct observation of the core of a big planet, such as Earth and Mars — especially if the probe vehicle can land on the parent body of the iron meteorite preserving a part of the silicate phase; we could then obtain information on metallic and silicate phases by exploring the boundaries between the metallic and silicate phases (Anders and Grevesse, 1989; Pernicka and Wasson, 1987). This would give us important information about the initial differentiation of the parent body.

Furthermore, the exploration of the parent body of an iron meteorite is important in resource exploration. The bulk compositions of iron meteorites contain higher abundances of siderophile elements ($20 \times$ CI) and platino elements (10–$100 \times$ CI) than those of chondritic materials. From the viewpoint of space resources, the exploration of M-type asteroids is important for Mankind's future.

7.5.3.2.4 M-Type asteroid

M-type asteroids have a flat reflectance spectrum. The largest M-type asteroid is 16 Psyche, and it appears to be metallic. 21 Lutetia was the first M-type asteroid to be imaged by the Rosetta space probe (Coradini *et al.*, 2011).

M-type asteroids are thought to be either naked cores or cores with a part of their mantle still intact. The widely accepted idea for the origins/creation of a naked core is that one or more giant collisions stripped the body of its silicate crust and mantle. The surviving metal core may have been molten before and after the stripping off of the mantle material. A melted core may produce a core dynamo. Differentiation in planetary formation is a fundamental process in shaping many asteroids and all terrestrial planets. A direct exploration of the core greatly enhances our understanding of the process. At the time of writing, it was announced that the Psyche Mission, an orbiter mission that seeks to achieve this by studying asteroid 16 Psyche, had been selected for the 13th Discovery mission.

Recent seismic observation and high-pressure experiments suggest that Earth's outer core contains ~8 wt% of light elements such as S, O, Si, C, and H, and that large-scale dynamics are closely associated with the superplumes originating at MCB, where light elements are carried away from the outer core.

The investigation of M-type asteroids is thought to help our understanding of Earth's dynamics, as well as provide possible answers for the sustainable maintenance of human activity in the future. Because asteroids and NEOs provide massive storages of valuable resources; the studying of NEOs' physical, mineralogical, chemical and biological properties is therefore of utmost scientific interest. Other reasons to visit NEOs are that they provide an opportunity to make space exploration sustainable — after all, amongst asteroids are NEOs with orbits very close to Earth's orbit — and to develop and establish a new branch of the human activity in space. These scientific and economic interests drive us to explore NEOs (Abell *et al.*, 2009; Gerlach, 2005). From the points-of-view of science and space resources, asteroid exploration is likely to grow in importance in the future. At present, M-type asteroids among NEOs have, unfortunately, not yet been identified. Since M-types are one of the three basic and major asteroid types in early classifications, and the number of NEOs larger than 140 m in size has increased ~1700 per year in 2015, it is thus safe to conclude that small M-type NEOs are expected to be discovered in the near future.

7.5.3.3 *Scientific objectives and goals*

The lack of direct samples from terrestrial planets' cores hampers the understanding of core formation processes. Many iron meteorites are thought to be fragments of metal cores from asteroidal parent bodies (Chabot and Haack, 2006; Haack and McCoy, 2005). But the spatial scale of this type of observation is quite limited. M-type asteroids, such as 16 Psyche, have been identified as Fe- and Ni-rich bodies based on IR spectra and radar observations, and close spectral matches with iron meteorites indicate that some M-type asteroids are composed mostly of planetary cores (Ockert-Bell *et al.*, 2010). Observations of a fly-by, rendezvous, or orbital mission around the asteroids has the potential to (1) provide independent constraints on models of core crystallization patterns based on iron meteorites, (2) characterize the relationship between core and mantle of a small differentiated body, (3) provide a mineralogical and chemical data set for better understanding of terrestrial planet cores, (4) reveal the major alloying elements in the iron metal core, (5) reveal the key characteristics of a geologic surface and its global topography (geology of metal bodies is new field), and (6) reveal how craters on a metal body differ from those on silicate rock (Ockert-Bell *et al.*, 2010).

Chemical and petrologic studies of iron meteorites show that their several groups formed by fractional crystallization in metallic liquids and originate from the cores of asteroidal parent bodies (Chabot and Haack, 2006; Wasson, 1999). Imaging of an asteroidal core in an M-type asteroid can provide compositional data over a spatial scale that can address core crystallization models. Characterizing the bulk Fe and Ni contents provides good diagnostic information on the thermal evolution of the core. And the measurement of light element (H, C, O, Si, S) contents of the metals leads us toward a better understanding of how light elements partition into cores during differentiation. Direct imaging of a significant portion of an asteroid core will provide some perspective for interpreting Earth's core. Moreover, a core/mantle boundary would offer a new example of a metal/silicate boundary to contrast with the high-pressure D layer on Earth.

With the use of a miniature deep space 50-kg-class microsatellite with an ion engine, we can easily and quickly reach, observe, and investigate NEOs (Funase *et al.*, 2014; Koizumi *et al.*, 2012; 2014; Ozaki *et al.*, 2014). Rendezvous and/or fly-by missions for M-type asteroids using microsatellites carrying nuclear spectrometers as their science payloads would be an attractive and interesting next step for deep space exploration using microsatellites.

7.5.4 *Nuclear spectrometer for NEA missions*

We propose the use of GNSs (gamma-ray and neutron spectrometers) on a 50-kg-class deep space microsatellite with a small ion engine for the investigation of M-type asteroids. Such a miniature deep space mission would visit NEOs (Hasebe *et al.*, 2014; 2015). As described earlier, GNS is already widely applicable to any NEO mission.

The GNS consists of the gamma-ray sensor (GS), the NS, and common electronics (CE). The GNS measures the surface abundances of major elements over the whole surface of the bodies, metallic regions, and stony regions of the asteroids. The observation goal of the GS is to provide the surface abundance of major elements over the whole surface, particularly of elements whose abundance is thought to be indicative of their origin (Boynton *et al.*, 2004; Feldman *et al.*, 2004a,b; Goldsten *et al.*, 2007; Harrington *et al.*, 1974; Hasebe *et al.*, 2008; Kim and Hasebe, 2012; Prettyman *et al.*, 2011; 2015a; Reedy, 1978). The NS measures the fluxes of thermal, epithermal, and fast neutrons emitted from the surface of the asteroids. These fluxes are so sensitive to the H concentration in the subsurface material (Feldman *et al.*, 2004a,b; Goldsten *et al.*, 2007; Kim and Hasebe, 2012; Prettyman, 2011; 2015a) that the NS is a key instrument to draw a strong constraint on whether the planetary body is primordial or differentiated.

7.5.4.1 *Gamma-ray sensor*

The gamma-ray sensor (GS) consists of an HPGe detector as a main detector and a thin plastic scintillator as anticoincidence detector (see Fig. 7.13a). The HPGe is an n-type, coaxial cylindrical Ge crystal with a volume of ~200 cm^3. The crystal is clamped in a hermetically sealed Al capsule. The HPGe crystal is cooled to an operating temperature in the 80–90 K range by a mechanical cryocooler rigidly attached to a base plate. The GS measures gamma rays with energies from 200 keV to 10 MeV. The energy resolution of HPGe is 0.4% or better at 1332 MeV, which is superior to those of the unit's scintillators. The anticoincidence counter in the GS greatly reduces background counts produced by abundant GCR charged particles.

The mechanical cryocooler is the Ricor K508 unit, which has space mission heritage from the Messenger GRS (Goldsten *et al.*, 2007). This cooler provides a nominal cooling capacity of 0.5 W at a cold tip temperature of 85 K for an ambient temperature of 20°C and 13 W of input power. The cooler's mass is ~450 g, including its motor drive electronics. The waste heat from the cooler compressor is emitted into a cold space.

A multilayer insulator surrounds the housing and is attached to the inner edges of the passive radiator.

7.5.4.2 *Neutron spectrometer*

A neutron spectrometer (NS) measures thermal, epithermal, and fast neutrons, whose energy ranges are less than 1 eV, from 1 eV to 500 keV, and from 500 keV to 7 MeV, respectively. The NS is composed of a 4-mm-thick ^{6}Li-enriched lithium glass scintillator, a BLP scintillator with a 4″ diameter and 3″ length, and a 2″ diameter photomultiplier tube (see Fig. 7.13b). The sidewall and front side of the plastic scintillator are wrapped with a 0.25-mm-thick Gd foil to absorb thermal neutrons. These scintillators are surrounded by a plastic scintillator that acts as a veto counter of GCR particles. The Li-glass scintillator detects nearly all the incident thermal neutrons via the ^{6}Li(n, α)^{3}H reaction. Thermal and epithermal neutrons are also detected by the ^{10}B(n, $\alpha\gamma$)^{7}Li reaction monitored by the borated plastic scintillator. Thermal neutrons can be filtered out by wrapping the borated plastic scintillator in a Gd foil, which strongly absorbs neutrons with energies below about 0.5 eV. Thus, the combination of a bare and Gd-covered scintillator can be used to separately measure contributions from thermal and epithermal neutrons. Fast neutrons ($E_n \geq$ 500 keV) can be detected by a prompt pulse from proton recoils followed a short time later by a second pulse produced by the neutron capture of the moderated neutron by ^{10}B. Thus, the delayed coincidence of a double pulse can be used to measure fast and epithermal neutron events separately.

The GS's mass and power consumption are 3.8 kg and 9 W, respectively, whereas those of the NS are 2.7 kg and 2 W and those for the CE are 1.5 kg and 3 W, respectively. The GNS's total mass and power are 8.0 kg and 14 W, respectively. The GNS's outstanding features are a high-resolution gamma-ray energy, being lightweight, and its low power consumption.

7.5.4.3 *Characteristics of GNS*

7.5.4.3.1 Summary

The gamma-ray and neutron spectrometer (GNS) will allow researchers globally (1) to characterize the broad geochemical regions of the asteroid

(a)

(b)

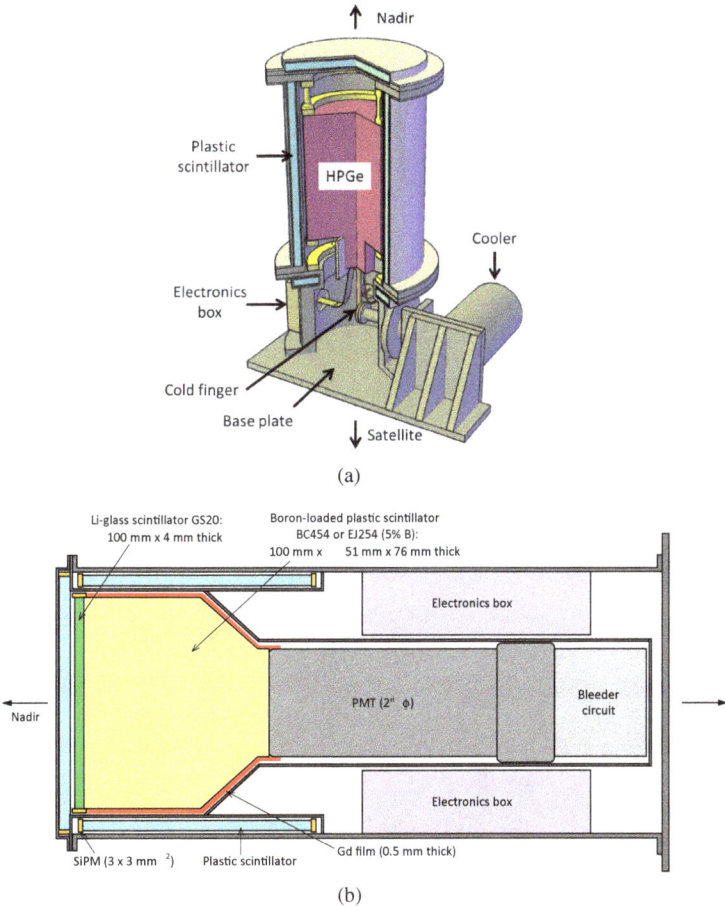

Figure 7.13:   Schematic drawing of a (a) gamma-ray sensor and (b) neutron sensor.

surface in spatially resolved measurements of Fe, Ni, Si, K, S, Al, Ca, Th, and U concentrations, as well as measure H and C; (2) to study geological features of formation and surface action of the NEA body; (3) to study the morphology of the impact craters; (4) to study the relation between M-type asteroids and Earth's core; and (5) to collect basic information on elements in asteroids among NEAs, which are expected to have high concentrations of heavy elements as Fe, Co, Ni, and rare metals such as those from the platinum group.

The technological and scientific challenges of NEO expeditions will provide good opportunities for us to collaborate with international partners from hardware and logistical services industries. A combined robotic and human NEO exploration in the future that is not solely justified by scientific benefits is certain to deliver surprising discoveries.

Chapter 8

Nuclear Planetary Science as a Tool to Study Life in the Universe

8.1 Introduction

Research on the origin and subsequent evolution of life on Earth may be extended to investigate life beyond the solar system. Since most of the events associated with the origin and subsequent evolution of life on Earth may be thought to apply universally to life yet to be found in cosmic space, it seems necessary for us to study life on Earth and then use the scientific results obtained to search for life in outer space.

This chapter briefly reviews scientific research conducted on subjects related to nuclear planetology or nuclear planetary science, while concentrating our efforts on life on Earth.

8.2 Characteristics of Life on the Present-Day Earth

As we know, every life form found on Earth seems to make use of the elements abundantly found on it. From these elements, every life-form chemically synthesizes organic molecules such as nucleotides, ribosome, RNA, and DNA, which are identified as essential for sustaining it. These chemical substances seem to have been processed during the evolution of life on Earth for 3.8 billion years or so.

It should be noted that the helicity of all lives ranging from plants to animals is left-handed. A possible cause of this uniqueness may have been related to the influence of the terrestrial magnetic field in the early period of their evolution. This magnetic field accidently originated on Earth about 3.8 billion years ago and strongly influenced the development of all life-forms (now theorized to be archaea; single-cell organisms) in the early period of the evolution. Archaea are now considered the precursor of both plants and animals.

8.3 Possible Origin of Life on Earth and Other Planets

The result of extensive efforts by the researchers engaged in the research on the origin and evolution of Earth, the epoch of when the first life-form originated on Earth has recently been identified as being about 3.8 billion years ago, just after the geomagnetic field was formed inside Earth. Since the origin of the first life-form on Earth, a variety of plants and animals have evolved. It seems that the first life on Earth was made of a creature consisting of just one cell — a form that might be similar to the species classified as archaea.

It was during this age, estimated to span as long as 3.8 billion years, that the life-forms currently seen on Earth were evolved. Thus, the fundamental nature of every life-form found on present-day Earth is based on the chemical substances described in Chapter 2. For now, there is no hint of the existence of any kind of any similar forms of life on planets other than Earth.

8.4 Search for Planets and their Associated Life in the Galaxy

The attempt to detect the planets and planetary systems beyond the solar system was made by the spacecraft called Kepler. It was launched in 2009 and continued to search for exoplanets for about two years, until its parts that detected exoplanets were accidentally damaged. Despite this, the attempt was considered successful because many exoplanets were discovered in the narrow space of about 100 square degrees surrounded by the two constellations, Lyra and Cygnus, in the direction of rotating motion of the Sun in the Milky Way.

According to the observed results, the number of planets and planetary systems estimated to be searched is about 10 times more than those of stars in the galaxy. Among them, more than 3000 planets, being not so far away from Earth, have water on their surfaces. So some of them may have environments conducive for the development of life. After all, it is well-established that water is considered necessary for life like those on Earth to originate and evolve on planets.

8.5 Nuclear Planetary Science as a Tool to Study Life in the Outer Space

A nuclear cosmochemical study should be conducted while searching for a suitable condition that makes it possible for any kind of life to originate and evolve in planetary environments. Various methods and techniques are used to investigate the physical processes taking place in the Earth's environment. With the radiochemical method used for analyzing ^{14}C content in tree rings and other materials, it is possible to find what chemical process occurred in the terrestrial environment for up to a hundred thousand years.

Some chemical substances such as NO_3 and CO_2 are useful for investigating what processes were taking place in the terrestrial environment during the evolution of Earth. Fossilized chemical substances with remains containing RNA, DNA and other organic substances could be used to look for the evolutionary processes in the atmosphere and sea during Earth's early period of evolution.

Various physical and chemical methods to analyze how Earth's atmosphere and oceans evolved have been used for the past 20 years or so. A new branch of science, named Nuclear Planetology or Nuclear Planetary Science, gives us information about various materials; contributing to the study of the evolutionary trends of various live-forms on Earth.

In regard to life sciences, it is important to understand how various amino acids such as glycine, leucine, and many others were chemically synthesized during the early periods of Earth's evolution. In order to find these synthetic mechanisms out, various experimental studies on gamma rays, ultraviolet rays, cosmic rays, and other high-energy beams of electrons with energies of 5 MeV or so, have been carried out. According to the results of the experiments performed, various amino acids necessary for life on Earth were chemically produced. These include glycine, alanine, aspartic acid, asparagines, and glutamic acid.

The amino acids just described are essentially important for the early evolution of life on Earth. Thus, nuclear cosmochemical research seems substantially important in the realm of Nuclear Planetology.

8.6 Some Specific Topics Associated with the Search for Life

Currently, except for life on Earth, we have no clues indicating the existence of any life-forms on other planets in the solar system, though some debris possibly related to primitive life on the surface of Mars seem to exist. The finding was made possible due to the exploration of the Mars rover, Curiosity. In 2015, before Curosity, the Mars Reconnaissance Orbiter (MRO) found evidence of the existence of historical water flows on Mars.

Beyond the solar system, more than one thousand planets have been detected by the Kepler mission. On the basis of the analyses of these planets in order to understand their relation with their parent stars. It is estimated that there are as many as 10 times more of such planets than there are parent stars. Among these planets, more than a thousand planets have water molecules content in their chemical substances. So some of these planets may have developed environments in which some life-forms might be in phases of evolution. Within ten years or so from now, some possible clue about the existence of life on those planets could hopefully be found.

In searching for such forms of life on the outer planets, the so-called exoplanets, it appears appropriate to recall some of the previous projects that searched for intelligent life in extraterrestrial space. The first attempt to search for such a life was initiated by Frank Donald Drake in 1960. He tried to find evidence of radio signals sent by intelligent life living on planets circling stars similar to the Sun. His research went on for about three months at the National Radio Astronomy Observatory located in Green Bank, West Virginia, USA, but no credible results were found.

The radio communication experiment was successfully performed in the middle of 1930. In early 1982, radio signals were sent out from this observatory to the stellar conglomerate named M13, which is more than 10,000 light-years away. As is well known, the terrestrial atmosphere is mostly transparent to microwave radio signals used for communication worldwide. So, some of these signals could have leaked into outer space, beyond the solar system. They might have now reached more than 80 light-years away from Earth and could have been captured by civilized intelligent life-forms that could have evolved and developed into some form of civilization there.

Life, primitive or civilized, seems to be universal as it appears that life in any form may have evolved in the universe. The morphological form of life on other planets seems to be similar to life found on Earth, but their chemical and biological processes may not be similar to those of life found on Earth. So, it is necessary to initially abandon our current view of what constitutes as life on Earth in order to study life beyond the solar system. However, it should be mentioned that any life evolving in this universe should necessarily use the chemical elements that are abundantly found on Earth. This suggests that any possible physiological processes may be similar to any life-forms living on Earth. This may mean that our initial search for life in the universe must be directed to discover some celestial objects similar to our Earth.

Furthermore, it should be mentioned that the rapid progress in the life sciences in recent years, has been in the realm of molecular biology and other related subjects.

Based on the arguments presented in this book — e.g. the scientific and commercial incentives — nuclear planetary science is therefore conclusively useful for future progress in the life sciences.

Chapter 9

Concluding Remarks

Based on the reviews described in the preceding chapters, it seems appropriate to summarize the research results accomplished up to the present in the various research fields in and associated with the discipline of nuclear planetary science. For this reason, the results attainted in both experimental and theoretical researches have been extensively reviewed.

The research in this field (so far) is largely defined by the nuclear cosmo-chemical methods being applied to the investigation of the planet Earth and various celestial bodies in the solar system, including asteroids of various types. As mentioned earlier, this research field adopts the physical and chemical nuclear methods in its investigations of the nature and properties of celestial objects in the solar system, in the aim to solve the question of the origins of these objects and their evolution after they were formed. And the main objective of the field recently established as nuclear planetary science may be summarized as the direct exploration and investigation of various celestial bodies in the solar system.

Research results recently attained in several institutions around the world, including our own research, have been reviewed in Chapters 5 and 6. Readers may confirm and recognize the current situation in research related to nuclear planetology. Major advances in the research were summarized in Chapter 6, and various topics in the current research were also reviewed. Research results considered important were reviewed in some detail in Chapters 5 and 6.

As stated in the text of this treatise, this research field is still in the state of infancy. As the *avant-garde* of this field, we naturally feel that extensive effort should be concentrated on every associated research field with the aim to attain rapid progress in nuclear planetary science in the coming years. In some way, this book is a part of our efforts and contribution towards increasing attention on this new research field.

We seek to strengthen the networking of researchers in the various fields of planetary science (and other inter-related subjects) in Japan and other countries, by promoting exchanges between different partners, including those from outside Japan, as well as by providing support to planetary exploration missions. This boosts synergies and joint research projects, in particular, through working groups, workshops, the exchange of personnel joint observation campaigns, and the development of dedicated tools for the easy sharing of data and information.

To build our natural interest in space and space exploration, we have to develop specific outreach and communication activities aimed at increasing the awareness and understanding of citizens, especially children and young people, of the results of planetary observation and space exploration programs. This may also help to attract budding scientists to this important field of research.

References

Abell, P. A., Korsmeyer. D. J., Landis, R. R., *et al.* (2009). Scientific exploration of near-Earth objects via the orion crew exploration vehicle, *Meteor. Planet. Sci.*, **44**, 1825–1836.

Adler, I., Trombka, J., Gerard, J., *et al.* (1972a). X-ray fluorescence experiment, NASA SP-289, *Appollo15 Preliminary Science Report*, pp. 17-1–17-17.

Adler, I., Trombka, J., Gerard, J., *et al.* (1972b). Apollo 15 geochemical X-ray fluorescence experiment: Preliminary report, *Science*, **28**, 436–440.

Adler, I. and Trombka, J. (1977). Orbital chemistry — Lunar surface analysis from the X-ray and gamma ray remote sensing experiments, *Phys. Chem. Earth*, **10**, 17–43.

Akkurt, H., Groves, J., Trombka, J., *et al.* (2005). Pulsed neutron generator system for astrobiological and geochemical exploration of planetary bodies, *Nucl. Instr. and Meth. B*, **241**, 232–237.

Alexander, C. M., *et al.* (2012). The provenances of asteroids, and their contributions to the volatile inventories of the terrestrial planets, *Science*, **337**, 721–723.

Anders, E. and Grevesse, N. (1989). Abundances of the elements: Meteoric and solar, *Geochimica Cosmochimica Acta*, **53**, 197–214.

Anderson, B. J., Acuña, M. H., Korth, H., *et al.* (2008). The structure of mercury's magnetic field from MESSENGER's first flyby, *Science*, **321**, 82–85.

Araki, H., Tazawa, S., Noda, H., *et al.* (2009). Lunar global shape and polar topography derived from Kaguya-LALT laser Altimetry, *Science*, **323**, 897–900.

Arnold, J. R., Honda, M. and Lal, D. (1961). Record of cosmic-ray intensity in the meteorites, *J. Geophys. Res.*, **66**, 3519–3531.

Arnold, J. R., Metzger, A. E., Anderson, E. C. and Van Dilla, M. A. (1962). Gamma rays in space, Ranger 3, *J. Geophy. Res.*, **67**, 4878–4880.

Bandfield, J. L., Hamilton, V. E., Christensen, P. R. (2000). A global view of martian surface compositions from MGS-TES. *Science*, **287**, 1626–1630.

Banerjee, D. and Gasnalt, O. (2008). Hard X rays and low-energy gamma rays from the Moon: Dependence of the ontinuum on the regolith composition and the solar activity, *J. Geophys. Res.*, **113**, E07004.

Basilevsky, A. T., Lorenz, C. A., Shingareva, T. V., *et al.* (2014). The surface geology and geomorphology of Phobos, *Planet Space Sci,* **102**, 95–118.

Bauer, C. A. (1948). The absorption of cosmic radiation in meteorites, *Phys. Rev.,* **74**, 225–226.

Bielefeld, M. J., Reedy, R. C., Metzger, A. E., *et al.* (1976). Surface chemistry of selected lunar regions, *Proc LPSC 7th*, pp. 2661–2676.

Binzel, R. P. and Xu, S. (1993). Chips off of asteroid 4 vesta: Evidence for the parent body of basaltic achondrite meteorites, *Science*, **260**, 186–191.

Binzel, R. P., Gaffcy. M. J., Thomas. P. C., *et al.* (1997). Geologic Mapping of Vesta from 1994 Hubble Space Telescope Images, *Icarus*, **128**, 95–103.

Birch, F. (1964). Density and composition of mantle and core, *J. Geophys. Res.*, **69**, 4377–4388.

Boynton, W. V., Feldman, W. C., Squyres, S. W., *et al.* (2002). Distribution of Hydrogen in the near surface of mars: Evidence for subsurface ice deposits, *Science*, **297**, 81–85.

Boynton, W. V., Feldman, W. C., Mitrofanov, I. G., *et al.* (2004). The mars odyssey gamma-ray spectrometer instrument suite, *Space Sci. Rev.*, **110**, 37–83.

Boynton, W.V., Taylor, G. J., Evans, L. G., *et al.* (2007). Concentration of H, Si, Cl, K, Fe, and Th in the low- and mid-latitude regions of Mars, *J. Geophys. Res.*, **112**, E12S99.

Brückner, J. and Masarik, J. (1997). Planetary gamma-ray spectroscopy of the surface of Mercury, *J. Planet. Space. Sci.*, **45**, 39–48.

Brückner, J., Reedy R. C., Englert, P. A. J., *et al.* (2011). Experimental simulations of planetary gamma-ray spectroscopy using thick targets irradiated by protons, *Nucl. Instr. and Meth. B*, **269**, 2630–2640.

Buffett, B. A. (2000). Earth's Core and the Geodynamo, *Science*, **288**, 2007–2012.

Buzhan, P., Dolgosheina, B., Filatov, L., *et al.* (2003). Silicon photomultiplier and its possible applications, *Nucl. Instr. and Meth. A*, **504**, 48–52.

BVSP (Basaltic Volcanism Study Project 1981, Basaltic Volcanism on the Terrestrial Planets, Pergamon, New York. 1286).

Camp, A., Vargas, A., and Fernández-Varea, J. M. (2016). Determination of $LaBr_3(Ce)$ internal background using a HPGe detector and Monte Carlo simulations, *Appl. Radiat. Isot.*, **109**, 512–517.

Capaccioni, F., Coradini, A., Filacchione, G., *et al.* (2015). The organic-rich surface of comet 67P/ Churyumov-Gerasimenko as seen by VIRTIS/Rosetta, *Science*, **347**, 389.

Chabot, N. L. and Haack, H., (2006). *Evolution of Asteroidal Cores*, Univ. of Arizona Press, Arizona, pp. 747–771.

Chin, G., Brylow, S., Foote, M., *et al.* (2007). Lunar reconnaissance orbiter overview: The instrument suite and mission, *Space Sci Rev.*, 129, pp. 391–419.

Citron, R. I., Genda. H. and Ida, S., (2015). Formation of phobos and deimos via a giant impact, *Icarus*, **252**, 334–338.

Colaprete, A., Schultz, P., Heldmann, J., *et al.* (2010). Detection of water in the LCROSS ejecta plume, *Science*, **330**, pp. 463–468.

Coradini, A., Capaccioni, F., Erard, S., *et al.* (2011). The surface composition and temperature of asteroid 21 lutetia as observed by rosetta/VIRTIS, *Science*, **334**, 492–494.

Craddock, R. A. (2011). Are phobos and deimos the result of a giant impact? *Icarus*, **211**, 1150–1161.

Cushing, J. A., Taylor, G. J., Norman, M. D. and Keil, K. (1999). The granulitic impactite suite: Impact melts and metamorphic breccias of the early lunar crust, *Meteor. Planet. Sci.*, **34**, 185–195.

d'Uston, C., Thocaven, J. J., Hasebe, N., *et al.* (2005). MANGA, a gamma-ray and neutron spectrometer for the BepiColombo mission, *Proc LPSC 36th*, 1873.

DeHon, R. A. and Waskom, J. D. (1976). Geologic structure of the eastern mare basins, *Proc LPSC 7th*, pp. 2729–2746.

Dohm, J. M., Baker, V. R., Boynton, W. V., *et al.* (2009). GRS evidence and the possibility of paleooceans on Mars, *Planet Space. Sci.*, **57**, 664–684.

Duffy, T. S. (2011). Probing the core's light elements, *Nature*, **479**, 480–481.

Elphic, R. C., Lawrence, D. J., Feldman, W. C., *et al.* (2000). Lunar rare earth element distribution and ramifications for FeO and TiO_2: Lunar Prospector neutron spectrometer observations, *J. Geophys. Res.*, **105**, 20333–20345.

Englert, P., Reedy, R. C. and Arnold, J. R. (1987). Thick-target bombardments with high energy charged particles: Experimental improvements and spatial distribution of low-energy secondary neutrons, *Nucl. Instr. and Meth. A*, **262**, 496–502.

Evans, L. G., Starr, R. D., Brückner, J., *et al.* (2001). Elemental composition from gamma-ray spectroscopy of the NEAR-Shoemaker landing site on 433 Eros, *Meteor. Planet. Sci.*, **36**, 1639–1660.

Evans, L. G., Boynton, W. V., Reedy, R. C., *et al.* (2002). Background lines in the Mars Odyssey 2001 gamma-ray detector, *Proc SPIE*, **4784**, 31–44.

Evans, L. G., Reedy, R. C., Starr, R. D., *et al.* (2006). Analysis of gamma ray spectra measured by Mars Odyssey, *J. Geophys. Res.*, **111**, E03S04.

Evans, L. G. and Peplowski, P. N. (2012). Major-element abundances on the surface of Mercury: Results from the MESSENGER Gamma-Ray Spectrometer, *J. Geophys. Res.*, **117**, E00L07.

Fabian, U., Masarik, J., Brückner, J., *et al.* (1996). Thick target experiments and monte carlo calculations for planetary gamma ray spectrometer, LPSC 27th, pp. 347–348.

Feldman, W. C., Maurice, S., Binder, A. B., *et al.* (1998). Fluxes of fast and epithermal neutrons from lunar prospector: Evidence for water ice at the lunar poles, *Science*, **281**, 1496–1500.

Feldman, W. C., Barraclough B.L., Fuller, K.R., *et al.* (1999). The lunar prospector gamma-ray and neutron spectrometers, *Nucl. Instr. and Meth. A*, **422**, 562–566.

Feldman, W. C., Maurice, S., Lawrence. D. J., *et al.* (2001). Evidence for water ice near the lunar poles, *J. Geophys. Res.*, **106**, 23231–23251.

Feldman, W. C., Boynton, W. V., Tokar, R. L., *et al.* (2002). Global Distribution of Neutrons from Mars: Results from Mars Odyssey, *Science*, **297**, 75–78.

Feldman, W. C., Ahola, K., Barraclough, B. L., *et al.* (2004a). Gamma-Ray, Neutron, and Alpha-Particle Spectrometers for the Lunar Prospector mission, *J. Geophys. Res.*, **109**, E07S06.

Feldman, W. C., Prettyman, T. H., Maurice. S., *et al.* (2004b). Global distribution of near-surface hydrogen on Mars, *J. Geophys. Res.*, **109**, E09006.

Fink, D., Klein, J., Middleton, R., *et al.* (1998). ^{41}Ca, ^{26}Al, and ^{10}Be in lunar basalt 74275 and ^{10}Be in the double drive tube 74002/74001, *Geochim Cosmochim Acta.*, **62**, 2389–2402.

Fischer, E. M. and Pieters, C. M. (1996). Composition and exposure age of the Apollo 16 Cayley and Descartes regions from Clementine data: Normalizing the optical effects of space weathering, *J. Geophys. Res.*, **101**, 2225–2234.

Forni, O., Gasnault., Diez, B., *et al.* (2009). Independent component analysis of the gamma ray spectrometer data of SELENE (Kaguya), *IEEE Whispers09*, **1**.

Funase, R., Kawaguchi, J., Mori, O., *et al.* (2014). 50kg-class Deep Space Exploration Technology Demonstration Micro-spacecraft PROCYON, *Proc AIAA/USA Conf. 27th*, SSC14-VI-3.

Gellert, R., Rieder, R., Brückner, J., *et al.* (2006). Alpha particle X-ray spectrometer (APXS): Results from Gusev crater and calibration report, *J. Geophys. Res.*, **111**, E02S05.

Gerlach, C. L. (2005). Profitably Exploiting Near-Earth Object Resources, *Proc Inter. Space Develop. Conf.* National Space Society, Washington DC.

Gibson, E. K. (1977). Volatile elements, carbon, nitrogen, sulfur, sodium, potassium and rubidium in the lunar regolith, *Phys. Chem. Earth.*, **10**, 57–62.

Gillis, J. J. and Jolliff, B. L., (1999). Lateral and vertical heterogeneity of thorium in the Procellarum KREEP terraine, LPI contri. 980, Lunar nd Planetary Inst. Houston, Tex., pp. 18–19.

Gillis, J. J., Jolliff, B. L. and Korotev, R. L. (2004). Lunar surface geochemistry: Global concentrations of Th, K, and FeO as derived from lunar prospector and Clementine data, *Geochim Cosmochim Acta.*, **68**, 3791–3805.

Gladstone, G. R., Hurley, D. M., Retherford, K. D., *et al.* (2010). LRO-LAMP Observations of the LCROSS Impact Plume, *Science*, **330**, 472–476.

Goldsten, J. O., Rhodes, E. A., Boynton, W. V., *et al.* (2007). The MESSENGER gamma-ray and Neutron spectrometer, *Space Sci. Rev.*, **131**, 339–391.

Goswami, J.N., Banerjee, D., Bhandari, N., e*t al.* (2005). High energy X-γ ray spectrometer on the Chandrayaan-1 mission to the Moon, *J. Earth Syst.,* **114**, 733–738.

Goswami, J.N. and Annadurai, M. (2008). Chandrayaan-1 mission to the Moon, *Acta Astronautica,* **63**, 1215–1220.

Goswami, J.N. and Annadurai, M. (2009). Chandrayaan-1: India's first planetary science mission to the moon, *Curr. Sci.*, **114**, 486–491.

Grande, M., Kellett, B. J., Howe, C., *et al.* (2007). The D-CIXS X-ray spectrometer on the SMART-1 mission to the Moon — First results, *Planet. Space Sci.*, **55**, 494–502.

Haack, H. and McCoy, T. J. (2005). Iron and Stony-iron Meteorites, *Meteorites, Comets, and Planets: Treatise on Geochemistry*, **1**, 325–345.

Haines, E. L., Etchegaray-Ramirez, M. I. and Metzger, A. E. (1978). Thorium concentrations in the lunar surface. II — Deconvolution modeling and its application to the regions of Aristarchus and Mare Smythii, *Proc. Lunar Planet. Sci. Conf. 9th*, pp. 2985–3013.

Hand, E. (2015). Comet close-up reveals a world of surprises, *Science*, **347**, 358–359.

Harrington, T. M., Marshall, J. H., Arnold, J. R., *et al.* (1974). The Apollo Gamma-Ray Spectrometer, *Nucl. Instr. and Meth.*, **118**, 401–411.

Hasebe, N., Shibamura. E., Miyachi, T., *et al.* (2008). Gamma-ray spectrometer (GRS) for lunar polar orbiter SELENE, *Earth, Planets Space*, **60**, 299–312.

Hasebe, N., Shibamura, E., Miyachi, T., *et al.* (2009). First results of high performance ge gamma-ray spectrometer onboard lunar orbiter SELENE (KAGUYA), *J. Phys. Soc. Jpn.*, **78**, 18–25.

Hasebe, N., Karouji, Y., Okudaira, O., *et al.* (2010). Distributions of K, Th, U and Rare Earth Metal in Procellarum KREEP Terrane, *Proc International Symposium on Lunar Science* 84th, pp. 84–89.

Hasebe, N., Kusano, H., Nagaoka, H., *et al.* (2014). Nuclear Spectroscopic Approach to Study M-type Asteroides, *Proc ISLPS*, 3–5 June 2014, Macau.

Hasebe, N., Kusano, H., Nagaoka, H., *et al.* (2015). Nuclear Spectroscopic Approach to Study M-type Near-Earth-Asteroids on the Miniature Deep Space Satellite, *Proc ISTS* 30th, 2015-f-24.

Hasebe, N., Ohta, T., Amano, Y., *et al.* (2016). An Investigation of Elemental Composition of Martian Satellites by Gamma-ray and Neutron Spectrometer, *Proc J. Phys. Society (JPS) 11th*, 040006. https://doi.org/10.7566/JPSCP.11.040006.

Haskin, L. A. and Waren P. D. (1991). 'Lunar chemistry', in *Lunar Sourcebook: A Users Guide to the Moon*, Cambridge University Press, Cambridge, pp. 357–474.

Haskin, L. A., Gillis, J. J., Korotev, R. L., *et al.* (2000). The materials of the lunar procellarum KREEP Terrane: A synthesis of the data from geomorphological mapping, remote sensing, and sample analysis, *J. Geophys. Res.*, **105**, 20403–20415.

Hayne, P. O., Greenhagen, B. T., Foote, M. C., *et al.* (2010). Diviner lunar radiometer observations of the LCROSS impact, *Science*, **330**, 477–479.

Head, J. W. (1976). Lunar volcanism in space and time, *Rev. Geophys. Space Phys.*, **14**, 265–300.

Heiken, G. H., Vaniman, D. T., French, B. M., *et al.* (1991). 'The Lunar Regolith', in Heiken, G. H., Vaniman, D. T. and French, B. M. (eds.), *Lunar Source Book: A Users Guide to the Moon*, Cambridge Univ. Press, Cambridge, pp. 285–356.

Hiesinger, H. and Head, J.W. (2006). New views of lunar geoscience: An introduction and overview, *Rev. Mineral & Geochem.*, **60**, 1–81.

Huang, H., Fei, Y., Cai, L., *et al.* (2011). Evidence for an oxygen-depleted liquid outer core of the Earth, *Nature*, **479**, 513–516.

Ishida, M., Maruyama, S., Suetsugu, D., *et al.* (1999). Superplume project: towards a new view of whole earth dynamics, *Earth Planets Space*, **51**, 1–5.

Isozaki, Y. (2009). Illawarra Reversal: The fingerprint of a superplume that triggered Pangean breakup and the end-Guadalupian (Permian) mass extinction, *Gondowana Res.*, **15**, 421–432.

Izidoro, A., Hghighipour, N., Winter, O. C. and Tsuchida, M. (2014). Terrestrial planet formation in a protoplanetary disk with a local mass depletion: A successful scenario for the formation of mars, *Astrophys. J.*, **782**, 31–50.

Jansson, P. A. (1997). 'Modern constrained nonlinear methods', in *Deconvolution of Images and Spectra*, Elsevier, New York, pp. 107–181.

Jarosewich, E. (1990). Chemical analyses of meteorites: A compilation of stony and iron meteorite analyses, *Meteor. Planet Sci.*, **25**, 323–337.

Jolliff, B. L., Gillis, J. J., Haskin, L A., *et al.* (2000). Major lunar crustal terrranes: Surface expressions and crust-mantle origins, *J. Geophys. Res.*, **105**, 4197–4216.

Jolliff, B. L. and Gillis, J. J. (2005). South pole-aitken basin and the composition of the lunar crust, *Meteor. Planet. Sci.*, **40**, A77.

Kato, M., Sasaki, S., Tanaka, K., *et al.* (2008). The Japanese lunar mission SELENE: Science goals and present status, *Adv. Space Res.*, **42**, 294–300.

Kelly, N. J., Boynton, W. V., Kerry, K., *et al.* (2006). Seasonal polar carbon dioxide frost on Mars: CO2 mass and columnar thickness distribution, *J. Geophys. Res.*, **111**, E03S07.

Khan, A., Mosegaa, K. and Rasmussen, K. L. (2000). A new seismic velocity model for the moon from a monte carlo inversion of the apollo lunar seismic data, *J. Geophys. Res.*, **111**, 1591–1594.

Khan, A. and Mosegaard, K. (2002). An inquiry into the lunar interior: A nonlinear inversion of the Apollo lunar seismic data, *J. Geophys. Res.*, **107**, 3-1–3-23.

Kim, K. J., Sisterson J.M., Englert, P.A.J., *et al.* (2002). Experimental cross-sections for the production of ^{10}Be from natural carbon targets with 40.6 to 500 MeV protons, *Nucl. Instr. and Meth. B*, **196**, 239–244.

Kim, K. J., Graham, I.G., Masarik, J., *et al.* (2007). Numerical simulations with the MCNPX and LAHET Code Systems compared with direct measurement of neutron fluxes in terrestrial environments, *Nucl. Instr. and Meth. B*, **259**, 637–641.

Kim, K. J., Masarik, J., Reedy R. C., *et al.* (2010). Numerical simulations of production rates for ^{10}Be, ^{26}Al and ^{14}C in extraterrestrial matter using the MCNPX code, *Nucl. Instr. and Meth. B*, **268**, 1291–1294.

Kim, K. J. and Hasebe, N. (2012). Nuclear planetology: Especially concerning the moon and mars, *Res. Astron. Astrophys.*, **12**, 1313–1380.

Klingelhöfer, G., Brückner, J., d'Uston, C., *et al.* (2007). The Rosetta Alpha Particle X-ray Spectrometer (APXS), *Space Sci. Rev.*, **128**, 383–396.

Knoll, G. F. (1999). *Radiation Detection and Measurement*, 3rd edition. John Wiley and Sons, New Jersey.

Kobayashi, S., Hasebe, N., Shibamura, E., *et al.* (2010). Determining the absolute abundances of natural radioactive elements on the lunar surface by the Kaguya gamma-ray spectrometer, *Space Sci. Rev.*, **154**, 193–218.

Kobayashi, S., Karouji, Y., Morota, T., *et al.* (2012). Lunar farside Th distribution measured by Kaguya gamma-ray spectrometer, *Earth Planet. Sci. Lett.*, **337**, 10.

Koizumi, H., Komurasaki, K. and Arakawa, Y. (2012). Development of the Miniature Ion Propulsion System for 50 kg Small Spacecraft, *Proc AIAA/ASME/SAE/ASEE Joint Propulsion Conference & Exhibit 48th*, AIAA2012–3949.

Koizumi, H., Inagaki, T., Kasagi, Y. *et al.*, (2014). Unified Propulsion System to Explore Near-Earth Asteroids by a 50 kg Spacecraft, *Proc AIAA/USA Conf. on Small Satelites 28th*, SSC14-VI-6.

Korotev, R. L. (1998). Concentrations of radioactive elements in lunar materials, *J. Geophys. Res.*, **103**, 1691–1701.

Korotev, R. L., Jolliff, B. L., Zeigler, R. A., *et al.* (2003). Feldspathic lunar meteorites and their implications for compositional remote sensing of the lunar surface and the composition of the lunar crust, *Geochim. Cosmochim Acta*, **67**, 4895–4923.

Kusano, H., Oyama, Y., Naito, M., *et al.* (2014). Development of an x-ray generator using a pyroelectric crystal for x-ray fluorescence analysis on planetary landing missions, *Proc SPIE 9213, Hard X-Ray, Gamma-Ray, and Neutron Detector Physics XVI*, 921316. No paper.

Kusano, H., Hasebe, N., Nagaoka, H., *et al.* (2016). Current development status of an X-ray generator for X-ray fluorescence analysis on space mission, *Proc J. Phys. Society (JPS)*, **11**, 04005.

Lawrence, D. J., Feldman, W. C., Barraclough, B. L., *et al.* (1998). Global elemental maps of the moon: The lunar prospector gamma-ray spectrometer, *Science*, **281**, 1484–1489.

Lawrence, D. J., Feldman, W. C., Barraclough, B. L., *et al.* (2000). Thorium abundance on the lunar surface, *J. Geophys. Res.*, **105**, 307–331.

Lawrence, D. J., Feldman, W. C., Elphic, R. C., *et al.* (2002). Iron abundances on the lunar surface as measured by the lunar prospector gamma-ray and neutron spectrometers, *J. Geophys. Res.*, **107**, 5130, 13-1-13-26.

Lawrence, D. J., Puetter, R. C., Elphic, R. C., *et al.* (2007). Global spatial deconvolution of Lunar Prospector Th abundances, *J. Geophys. Res. Lett.*, **34**, L03201.

Lawrence, D. J., Feldman, W. C., Goldsten, J. O., *et al.* (2013). Evidence for water Ice near mercury's north pole from MESSENGER neutron spectrometer measurements, *Science*, **339**, 292–296.

Leya, I. and Michel, R. (2011). Cross sections for neutron-induced reactions up to 1.6 GeV for target elements relevant for cosmochemical, geochemical, and technological applications, *Nucl. Instr. and Meth. B*, **269**, 2487–2503.

Li, J.-Y., McFadden, L. A., Thomas, P. C., *et al.* (2010). Photometric mapping of Asteroid (4) Vesta's southern hemisphere with Hubble Space Telescope, *Icarus*, **208**, 238–251.

Lognonné, P. (2005). Planetary seismology, *Ann. Rev. Earth Planet. Sci.*, **33**, 571–604.

Lucy, P.G., Taylor, G.j., Malarel, E. (1995). Abundance and distribution of iron on the Moon, *Science*, **268**, 1150–1153.

Lucey, P. G. (1998). Model near-infrared optical constants of olivine and pyroxene as a function of iron content, *J. Geophys. Res.*, **103**, 1703–1713.

Luke, P. N. (1995). Unipolar charge sensing with coplanar electrodes-application to semiconductor detectors, *IEEE Trans. Nucl Sci.*, **42**, 207–213.

Ma, T., Chang, J., Zhang, N., *et al.* (2008). Gamma-ray detector on board lunar mission Chang'E-1, *Adv. Space Res.*, **42**, 347–349.

Ma, T., Chang, J., Zhang, N., *et al.* (2013). Gamma-ray spectrometer onboard Chang'E-2, *Nucl. Instr. and Meth. A*, **726**, 113–115.

Maruyama, S., Kumazawa, M. and Kawakami, S. (1994). Towards a new paradigm on the Earth's dynamics, *J. Geol. Soc. Jpn.*, **100**, 1–3.

Maruyama, S. (1994). Plume Techtonics, *J. Geol. Soc. Jpn.*, **100**, 24–49.

Masarik, J. and Reedy, R. C. (1996). Gamma ray production and transport in Mars, *J. Geophys. Res.*, **101**, 18,891–18,912.

McCord, T. B., Adams, J. B. and Johnson, T. V. (1970). Asteroid vesta: Spectral reflectivity and compositional implications, *Science*, **168**, 1445–1447.

McKay, D. S., Heiken, G., Basu, A., *et al.* (1991). 'The Lunar Regolith', in Heiken, G. H., Vaniman, D. T. and French, B. M. (eds.), *Lunar Sourcebook: A User's Guide to the Moon*, Cambridge Univ. Press, Cambridge, pp. 285–356.

McKinney, G.W., Lawrence, D. J., Prettyman, T. H., *et al.* (2006). MCNPX benchmark for cosmic ray interactions with the Moon, *J. Geophys. Res.*, **111**, E06004.

McSween, H. N., Binzel, R. P., De Sanctis, M. C., *et al.* (2013). Dawn; the Vesta–HED connection; and the geologic context for eucrites, diogenites, and howardites, *Meteorit. Planet. Sci.*, **48**, 2090–2104.

Metzger, A. E., Trombka, J. I., Peterson, L. E., *et al.* (1973). Lunar surface radioactivity: Preliminary results of the apollo 15 and apollo 16 gamma-ray spectrometer experiments, *Science*, **179**, 800–803.

Metzger, A. E., Haines, E. L., Parker, R. E., *et al.* (1977). Thorium concentrations in the lunar surface. I: Regional values and crustal content, *Proc. LSC 8th*, pp. 949–999.

Metzger, A. E. (1993). 'Composition of the Moon as Determined from Orbit by Gamma Ray Spectroscopy' in Pieters, C.P. and Englert, P.A.J. (eds), *Remote Geochemical Analysis,* Cambridge Univ. Press, Cambridge, pp. 341–365.

Michel, R., Dragovitsch, P., Englert, P., *et al.* (1986). On the depth dependence of spallation reactions in a spherical thick diorite target homogeneously irradiated by 600 MeV protons, *Nucl. Instr. and Meth. B*, **16**, 61–82.

Mitrofanov, I. G., Anfimov, D., Kozyrev, A., *et al.* (2002). Maps of subsurface hydrogen from the high energy neutron setector, Mars Odyssey, *Science*, **297**, 78–81.

Mitrofanov, I. G., Litvak, M. L., Kozyrev A. S. *et al.* (2009). Nuclear Instruments and Methods for Space Planetology: Recent Results and New Developments, LPSC 40th, [#1207].

Mitrofanov, I. G., Sanin, A. B., Boynton, W. V., *et al.* (2010). Hydrogen mapping of the Lunar South Pole using the LRO neutron detector experiment LEND, *Science*, **330**, 483–486.

Moszynski, M., Zalipska, J., Balcerzyk, M., *et al.* (2002). Intrinsic energy resolution of NaI(Tl), *Nucl. Instr. and Meth. A*, **484**, 259–269.

Naito, M., Hasebe, N., Yoshida, K., *et al.* (2016). Neutron Fluxes from Martian Satellites as a Function of Chemical Composition and Hydrogen Content, *Proc J. Phys. Society (JPS)*, **11**, 05003. https://doi.org/10.7566/JPSCP.11.050003.

Nakamura, Y. and Koyama. J. (1982). Seismic Q of the lunar upper mantle, *J. Geophys. Res.*, **87**, 4855–4861.

Naranjo, B., Gimzewski, J. K. and Putterman, S. (2005). Observation of nuclear fusion driven by a pyroelectric crystal, *Nature*, **434**, 1115–1117.

NASA, http://curator.jsc.nasa.gov/antmet/mmc/contents.cfm, 2/2/2016.

Norman, M. D., Borg, L. E., Nyquist, L. E. and Bogard, D. D. (2003). Chronology, geochemistry, and petrology of a ferroan noritic anorthosite clast from Descartes breccia 67215: Clues to the age, origin, structure, and impact history of the lunar crust, *Meteo. Planet. Sci.*, **38**, 645–661.

Ockert-Bell, M., Clark, B. E., Shepard, M. K., *et al.* (2010). The composition of M-type asteroids: Synthesis of spectroscopic and radar observations, *Icarus*, **210**, 674–692.

Ohtake, M., Matsunaga, T., Haruyama, J., *et al.* (2009). The global distribution of pure anorthosite on the Moon, *Nature*, **461**, 236–240.

Ozaki, N., Funase, R., Nakajima, N., *et al.* (2014). Preliminary Mission Design of PROCYON: A Micro Spacecraft to Asteroid, *24th Int. Symp. On Space Flight Dynamics*, Maryland USA., May 5–9, 2014.

Paige, D.A., Siegler, M. A., Zhang, J. A., *et al.* (2010). Diviner lunar radiometer observations of cold traps in the moon's South Polar region, *Science*, **330**, 479–482.

Paneth, F. A., Reasbeck, P. and Mayne, K. I. (1952). Helium 3 content and age of meteorites, *Geochim. Cosmochim. Acta.*, **2**, 300–303.

Papike, J. J., Ryder, G. and Shearer, C. K. (1998). Lunar samples, *Rev Mineral*, **36**, 5.1–5.234.

Parsons, A., Bodnarik, J., Evans, L., *et al.* (2011). Active neutron and gamma-ray instrumentation for in situ planetary science applications, *Nucl. Instr. and Meth. A*, **652**, 674–679.

Paul, R. L., Harris, L. J., Englert, P. A. J., *et al.* (1995). Production of cosmogenic nuclides in thick targets alpha bombardment. Part I — short-lived radioisotopes, *Nucl. Instr. and Meth. B*, **100**, 464–470.

Peplowski, P. N., Evans, L. G., Hauck, S. A., *et al.* (2011). Radioactive elements on mercury's surface from MESSENGER: Implications for the planet's formation and evolution, *Science*, **333**, 1850–1852.

Peplowski, P. N., Lawrence, D. J., Rhodes, E. A., *et al.* (2012a). Variations in the abundances of potassium and thorium on the surface of Mercury: Results from the MESSENGER Gamma-Ray Spectrometer, *J. Geophys. Res.*, **117**, E00L04.

Peplowski, P. N., Rhodes, E. A., Hamara, D. K., *et al.* (2012b). Aluminum abundance on the surface of Mercury: Application of a new background-reduction technique for the analysis of gamma-ray spectroscopy data, *J. Geophys. Res*, **117**, E00L10.

Peplowski, P. N., Lawrence, D. J., Prettyman, T. H., *et al.* (2013). Compositional variability on the surface of 4 Vesta revealed through GRaND measurements of high-energy gamma rays, *Meteor. Planet. Sci.*, **48**, 2252–2270.

Peplowski, P. N., Lawrence, D. J., Feldman, W. C., *et al.* (2015). Geochemical terranes of Mercury's northern hemisphere as revealed by MESSENGER neutron measurements, *Icarus*, **253**, 346–363.

Pernicka, E. and Wasson, J. T. (1987). Ru, Re, Os, Pt and Au in iron meteorites, *Geochim. Cosmochim. Acta.*, **51**, 1717–1726.

Pieters, C. M. and Englert, P. A. J. (1993). *Remote Geochmical Analysis: Elemental and Mineral Composition*, Cambridge University Press, Cambridge.

Pieters, C. M., Tompkins, S., Head, J. W. and Hess, P. C. (1997). Mineralogy of the Mafic Anomaly in the South Pole-Aitken Basin: Implications for excavation of the lunar mantle, *J. Geophys. Res. Lett*, **24**, 1903–1906.

Pieters, C. M., Head, J. W., Gaddis, L., *et al.* (2001). Rock types of South Pole-Aitken basin and extent of basaltic volcanism, *J. Geophys. Res. Lett.*, **106**, 28001–28022.

Pieters, C. M., Goswami, J. N., Clark, R. N., *et al.* (2009). Character and spatial distribution of OH/H_2O on the surface of the Moon seen by M^3 on Chandrayaan-1, *Science*, **326**, 568–572.

Pina, R. K. and Puetter, R. C. (1993). Bayesian image reconstruction: The pixon and optimal image modeling, *Publ. Astron. Soc. Pac.*, **105**, 630–637.

Prescott, J. R. and Narayan, G. H. (1969), Electron responses and intrinsic linewidths in NaI(TI) *Nucl. Instr. and Meth.*, **75**, 51–55.

Prettyman, T. H., Feldman, W. C., Ameduri, F. P., *et al.* (2003). Gamma-ray and neutron spectrometer for the dawn mission to 1 Ceres and Vesta, *IEEE-NS*, **50**, 1190–1197.

Prettyman, T. H., Hagerty, J. J., Elphic, R. C., *et al.* (2006). Elemental cpmposition of lunar surface: Analysis of gamma ray spectroscopy data from Lunar prospector, *J. Geophys. Res.*, **111**, E12007.

Prettyman, T. H., Feldman, W. C., McSween, H. M., *et al.* (2011). Dawn's gamma ray and neutron detector, *Space Sci Rev.*, **163**, 371–459.

Prettyman, T. H., Mittlefehldt, D. W., Yamashita, N., *et al.* (2012). Elemental mapping by dawn reveals exogenic H in Vesta's regolith, *Science*, **338**, 242–246.

Prettyman, T. H., Mittlefehldt, D. W., Yamashita, N., *et al.* (2013). Neutron absorption constraints on the composition of 4 Vesta, *Meteor. Planet. Sci.*, **48**, 2211–2236.

Prettyman, T. H. (2015a). in Spohn, T., Breuer, D. and Johnson, T., (eds.), *Encyclopedia of the Solar System*, 3rd Edition, 1161.

Prettyman, T. H., Yamashita, N., Reedy, R. C., *et al.* (2015b). Concentrations of potassium and thorium within Vesta's regolith, *Icarus*, **259**, 39–52.

Puetter, R. C., Gosnell, T. R., and Yahil, A. (2005). Digital image reconstruction: Deblurring and Denoising, *Ann. Rev. Astron. Astrophys.*, **43**, 139–194.

Reedy, R. C. and Arnold, J. R., (1972). Interaction of solar and galactic cosmic-ray particles with the moon, *J. Geophys. Res.*, **77**, 537–555.

Reedy, R. C., Arnold, J. R. and Trombka, J. L. (1973). Expected γ ray emission spectra from the lunar surface as a function of chemical composition, *J. Geophys. Res.*, **78**, 5847–5866.

Reedy, R. C. (1978). Planetary gamma ray spectroscopy, *Proc. LPSC 9th*, pp. 2961–2984.

Reedy, R. C., Masarik, J., Nishiizumi, K., *et al.* (1993). Cosmogenic-radionuclide profiles in Knyahinya: New measurements and models, *Proc LPSC 24th*, pp. 1195–1196.

Reedy, R. C. and Masarik, J. (1994). Cosmogenic-nuclide depth profiles in the lunar surface, *Proc LPSC 25th*, pp. 1119–1120.

Reedy, R. C. and Frankle, S. C. (2002). Prompt gamma rays from radiative capture of thermal neutrons by elements from hydrogen through zinc, *Atomic Data and Nuclear Data Tables*, **80**, 1–34.

Rhodes, E. A., Evans, L. G., Nittler, L. R., *et al.* (2011). Analysis of MESSENGER gamma-ray spectrometer data from the Mercury flybys, *Planet. Space Sci.*, **59**, 1829–1841.

Rieder, R., Wänke, H., Economou, T. and Turkevich, A. (1997). Determination of the chemical composition of Martian soil and rocks: The alpha proton X-ray spectrometer, *J. Geophys. Res.*, **102**, 4027–4044.

Rieder, R., Gellert, R., Brückner, J., *et al.* (2003). The new Athena alpha particle X-ray spectrometer for the Mars exploration rovers, *J. Geophys. Res.*, **108**, E128066.

Rosenblatt, P. and Charnoz, S. (2012). On the formation of the Martian moons from a circum-Martian accretion disk, *Icarus*, **221**, 806–815.

Russell, C. T., Raymond, C. A., Coradini, A., *et al.* (2012). Dawn at Vesta: Testing the protoplanetary paradigm, *Science*, **336**, 684–686.

Sakai, E. (1987). Recent measurements on scintillator-photodetector systems, *IEEE-NS*, **34**, 418–422.

Sasaki, S., Tawara, H., Saito, K., *et al.* (2010). Ws values in several inorganic scintillation crystals for gamma rays, *IEEE-NS*, **57**, 1282–1286.

Schmedemann, N. Michael, G. G., Ivanov, B. A., *et al.* (2014). The age of phobos and its largest crater, stickney, *Planetary and Space Science*, **102**, 152–163.

Schrader, C. D. and Stinner, J. (1961). Remote analysis of surfaces by neutron-gamma-ray inelastic scattering technique, *J. Geophys. Res.*, **66**, 1951–1956.

Schrunk, D. G. Sharpe, B., Cooper, B. L. and Thangavelu, M. (2008). *The Moon Resources, Future Development and Settlement*, Springer-Praxis, Berlin.

Schultz, P. H., Hermalyn, B., Colaprete. A., *et al.* (2010). The LCROSS cratering experiment, *Science*, **330**, 468–472.

Seabury, E. H., Wharton, J. C., Caffrey, A. J. (2006). Response of a $LaBr_3(Ce)$ Detector to 2-11 MeV Gamma Rays, *Proc IEEE Nuclear Science Symposium*, INL/CON-06-11300.

Shah, K. S., Glodo, J., Higgins, W., *et al.* (2005). $CeBr_3$ Scintillators for gamma-ray spectroscopy, *IEEE NS*, **52**, 3157–3159.

Shibata, K., Iwamoto, O., Nakagawa, T., *et al.* (2011). JENDL-4.0: A New Library for Nuclear Science and Engineering, *J. Nucl. Sci. Technol.* **48**(1), 1–30.

Sisterson, J. M., Kim, K., Beverding, A., *et al.* (1997), Measurement of proton production cross sections of ^{10}Be and ^{26}Al from elements found in lunar rocks, *Nucl. Instr. and Meth. B*, **123**, 324–329.

Snyder, G. A., Taylor, L. A. (1993). Constraints on the genesis and evolution of the Moon's magma ocean and derivative cumulate sources as supported by lunar meteorites, *Proc. NIPR Symp. Antarct. Meteorites*, **6**, 246–267.

Spudis, P.D., Hood, L. L. (1992). Geological and geophysical field investigations from a lunar base at Mare Smythii, *Proc 2nd conference on Lunar Bases and Space Activities of the 21st Century*, **1**, 163–174.

Spudis, P. D., Reisse, R. A., Gillis, J. I., *et al.* (1994). Ancient multiring basins on the Moon revealed by clementine laser altimetry, *Science*, **266**, 1848–1851.

Spudis, P. D. and Bussey, D. B., (1997). *Clementine Explores the Moon*, 2nd edition, LPI contri. 929, Lunar and Planetary Institute, Houston, Texas.

Taylor, S. R. and Jakes, P. (1974). The geochemical evolution of the moon, *Proc. LPSC 5th*, **2**, pp. 1287–1305.

Taylor, S. R. (1975). *Lunar Science: A Post-Apollo View: Scientific Results and Insights from the Lunar Samples*, Pergamon Press, Oxford.

Taylor, S. R. (1982). Planetary Science: A Lunar Perspective, 481, Lunar and Planet. Inst., *Tex.*

Taylor, G. J., Warren, P., Ryder, G., *et al.* (1991). Lunar rocks, in Heiken, G. H., Vaniman, D. T. and French, B. M. (eds.), *Lunar Sourcebook: A Users Guide to the Moon, Cambridge Univ. Press*, Cambridge, pp. 183–284.

Taylor, G. J., Stopar, J. D., Boynton, W. V., *et al.* (2006). Variations in K/Th on Mars, *J. Geophys. Res.*, **111**, E03S06.

Thomas, N., Stelter, R., Ivanov, A., *et al.* (2011). Spectral heterogeneity on Phobos and Deimos: HiRISE observations and comparisons to Mars Pathfinder results, *Planet Space Sci.*, **59**, 1281–1292.

Toksöz, M. N., Press, F., Dainty, A., *et al.* (1972). Structure, composition, and properties of lunar crust, *Proc. LPSC 3th*, pp. 2527–2544.

Toksöz, M. N., Dainty, A. M., Solomon, S. C., *et al.* (1974). Structure of the Moon, *Rev. Geophys.*, **12**, 539–567.

Trombka, J. I., Arnold, J. R., Reedy, R. C., *et al.* (1973). Some correlations between measurements by the Apollo gamma-ray spectrometer and other lunar observations, *Proc. LSC* 4th, pp. 2847–2853.

Trombka, J. I., Arnold, J. R., Reedy, R. C., *et al.* (1973). Proc. 4th lunar planet. science Conf., *Geochim. Cosmochim. Acta*, **3**, 2847–.

Trombka, J. I., Schmadebeck, R. L., Bielefeld, M. J., *et al.* (1979). 'Analytical methods in determining elemental composition from the Apollo X-ray and gamma-ray spsctrometer data', in (eds.) *Computers in Activation Analysis and gamma-ray spectroscopy Radiation Detection and Measurement*, 3rd edition. John Wiley and Sons, New Jersey.

Trombka, J. I., Nittler, L. R., Starr, R. D., *et al.* (2001). The NEAR-shoemaker x-ray/gamma-ray spectrometer experiment: Overview and lessons learned, *Meteor. Planet. Sci.*, **36**, 1605–1616.

Usui, T., McSween, H. Y., Mittlefehldt, D. W., Prettyman, T. H., *et al.* (2010). K-Th-Ti systematics and new three-component mixing model of HED meteorites: Prospective study for interpretation of gamma-ray and neutron spectra for the Dawn mission, *Meteor. Planet. Sci.*, **45**, 1170–1190.

Vadawale, S. V., Sreekumar, P., Acharya, Y. B., *et al.* (2014). Hard X-ray continuum from lunar surface: Results from high Eenergy X-ray spectrometer (HEX) onboard Chandrayaan-1, *Adv. Space Res.*, **54**, 2041–2049.

van Loef, E. V., Dorenbos, P., van Eijk, C. W. E., *et al.* (2002). Scintillation properties of $LaBr_3:Ce^{3+}$ crystals: Fast, efficient and high-energy-resolution scintillators, *Nucl. Instr. and Meth. A*, **486**, 254–258.

van Loef, E. V., Wilson, C. M., Cherepy, N. J., *et al.* (2009). Crystal growth and scintillation properties of strontium iodide scintillators, *IEEE NS*, **56**, 869–872.

Vinogradov, A. P., Surkov, Y. A., Chernov, G. M., *et al.* (1966). Measurements of gamma radiation of the lunar surface by the Luna-10 space station, *Cosmic Research*, **4**, 751.

Warren, P. H. and Wasson, J. T. (1979). The origin of KREEP, *Rev. Geophys. Space Phys.*, **17**, 73–88.

Warren, P. H. (1985). The magma ocean concept and lunar evolution, *Ann. Rev. Earth Planet. Sci.*, **13**, 201–240.

Wasson, J. T. (1999). Trapped melt in IIIAB irons; solid/liquid elemental partitioning during the fractionation of the IIIAB magma, *Geochim. Cosmochim. Acta.*, **63**, 2875–2889.

Wieczorek, M. A. and Phillips, R. J. (1999). Lunar multiring basins and the cratering process, *Icarus*, **139**, 246–259.

Wieczorek, M. A. and Phillips, R. J. (2000). The "Procellarum KREEP Terrane": Implications for mare volcanism and lunar evolution, *J. Geophys. Res.*, **105**, 20,417–20,430.

Wieczorek, M. A. and Zuber, M. T. (2004). Thickness of the Martian crust: Improved constraints from geoid-to-topography ratios, *J. Geophys. Res.*, **106**, E01009.

Wieczorek, M.A., Jolliff, B. L. Khan, M., *et al.* (2006). The constitution and structure of the lunar interior, *Rev. Mineral. & Geochem.*, **60**, 221–634.

Wilhelms, D.E., John, F. and Trask, N. J. (1987). The geologic history of the Moon, *U.S. Geol. Surv. Proc Paper,* p. 1348.

Witasse, O., Duxbury, T., Chicarro, A., *et al.* (2014). Mars express investigations of phobos and deimos, *Planet. Space Sci.*, **102**, 18–34.

Wootton, A. (2006). Earth's Inner Fort Knox, *Discover*, **27**, 18.

Yamashita, N., Hasebe, N. Miyachi, T. M. *et al.* (2006). Energy spectra of prompt gamma rays from Al and Fe thick targets irradiated by helium and proton beams: Concerning planetary gamma-ray spectroscopy, *J. Phys. Soc. Japan*, **75**, 054201–.

Yamashita, N., Hasebe, N., Miyachi, T., *et al.* (2008). Complexities of gamma-ray line intensities from the lunar surface, *Earth Planets Space*, **60**, 313–319.

Yamashita, N., Hasebe, N., Reedy, R. C., *et al.* (2010). Uranium on the Moon: Global distribution and U/Th ratio, *J. Geophys. Res. Lett.*, **37**, L10201.

Yamashita, N., Gasnault, O., Forni, O., *et al.* (2012). The global distribution of calcium on the Moon: Implications for high-Ca pyroxene in the eastern mare region, *Earth Planet. Sci. Lett.*, **353–354**, 93–98.

Yamashita, N., Prettyman, T. H., Mittlefehldt, D. W., *et al.* (2013). Distribution of iron on Vesta, *Meteor. Planet. Sci.*, **48**, 2237–2251.

Yates, S. W., Filo, A. J., Cheng. C. Y. and Coope, D. F. (1978). Elemental analysis by gamma-ray detection following inelastic neutron scattering, *J. Raioanal. Chem.*, **46**, 343–355.

Yingst, R. A. and Head, J. W. (1997). Volumes of lunar lava ponds in South Pole-Aitken and Orientale Basins: Implications for eruption conditions, transport mechanisms, and magma source regions, *J. Geophys. Res.*, **102**, 10909–10931.

Yoshida, K., Naito, M., Hasebe, N., *et al.* (2016). Gamma-ray emission from the surface of martian satellites as a function of elemental composition, *Proc J. Phys. Society (JPS) 11th*, 040007. https://doi.org/10.7566/JPSCP.11.040007.

Yoshimori, M., Hirayama, H., Mori, S., *et al.* (2003). Be-7 nuclei produced by galactic cosmic rays and solar energetic particles in the earth's atmosphere, *Adv. Space Res.*, **32**, 2691–2696.

Yuen, D. A., Maruyama, S., Karato, S.-I., Windley, B. F. (Eds.), (2007). *Superplumes: Beyond Plate*, Springer, Netherlands.

Zhu, M-H., Chang, J., Xie, M., Fritz, J., Fernandes, V. A, Ip, W-H., Ma, T., Xe, A., (2015). The uniform K distribution of the mare deposits in the orientale basin: Insights from Chang'E-2 gamma-ray spectrometer, *Earth and Planetary Science Letters*, **418**, 172–181. http://dx.doi.org/10.1016/j.epsl.2014.11.009.

Zwittlinger, H. (1973). Neutron-capture gamma-ray activation analysis of refined steel, *J. Radioanalytical. Chem.*, **14**, 147–158.

Index